my **revisio**

D1382476

WJEC/Eduqas AS/A-level

GEOGRAPHY

Kevin Davis
Simon Oakes

HODDER
EDUCATION
AN HACHETTE UK COMPANY

Acknowledgements

Every effort has been made to trace all copyright holders, but if any have been inadvertently overlooked, the Publishers will be pleased to make the necessary arrangements at the first opportunity.

Although every effort has been made to ensure that website addresses are correct at time of going to press, Hodder Education cannot be held responsible for the content of any website mentioned in this book. It is sometimes possible to find a relocated web page by typing in the address of the home page for a website in the URL window of your browser.

Hachette UK's policy is to use papers that are natural, renewable and recyclable products and made from wood grown in sustainable forests. The logging and manufacturing processes are expected to conform to the environmental regulations of the country of origin.

Orders: please contact Bookpoint Ltd, 130 Park Drive, Milton Park, Abingdon, Oxon OX14 4SE. Telephone: (44) 01235 827827. Fax: (44) 01235 400401. Email education@bookpoint.co.uk Lines are open from 9 a.m. to 5 p.m., Monday to Saturday, with a 24-hour message answering service. You can also order through our website: www.hoddereducation.co.uk

ISBN: 978 1 5104 4768 4

First published in 2019 by

Hodder Education,
An Hachette UK Company
Carmelite House
50 Victoria Embankment
London EC4Y 0DZ

www.hoddereducation.co.uk

Impression number 10 9 8 7 6 5 4 3 2 1

Year   2023   2022   2021   2020   2019

Cover photo © Romolo Tavani – stock.adobe.com

Typeset in BemboStd 11/13 pts by Aptara, Inc.

Printed in Spain

A catalogue record for this title is available from the British Library.

Get the most from this book

Everyone has to decide his or her own revision strategy, but it is essential to review your work, learn it and test your understanding. These Revision Notes will help you to do that in a planned way, topic by topic. Use this book as the cornerstone of your revision and don't hesitate to write in it — personalise your notes and check your progress by ticking off each section as you revise.

Tick to track your progress

Use the revision planner on page 4 to plan your revision, topic by topic. Tick each box when you have:

- revised and understood a topic
- tested yourself
- practised the exam questions and gone online to check your answers and complete the quick quizzes

You can also keep track of your revision by ticking off each topic heading in the book. You may find it helpful to add your own notes as you work through each topic.

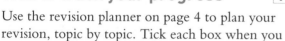

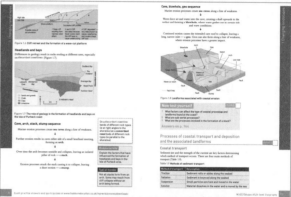

Features to help you succeed

Expert tips are given throughout the book to help you polish your exam technique in order to maximise your chances in the exam.

Typical mistakes

The authors identify the typical mistakes candidates make and explain how you can avoid them.

Now test yourself

These short, knowledge-based questions provide the first step in testing your learning. Answers are at the back of the book.

Definitions and key words

Clear, concise definitions of essential key terms are provided where they first appear.

Key words from the specification are highlighted in bold throughout the book.

Revision activities

These activities will help you to understand each topic in an interactive way.

Case study

These are named examples included to illustrate the key concepts as required by the specification.

Exam practice

Practice exam questions are provided for each topic. Use them to consolidate your revision and practise your exam skills.

Summaries

The summaries provide a quick-check bullet list for each topic.

Online

Go online to check your answers to the exam questions and try out the extra quick quizzes at **www.hoddereducation.co.uk/ myrevisionnotesdownloads**

My revision planner

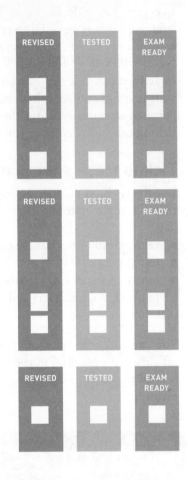

**Exam practice answers and quick quizzes at
www.hoddereducation.co.uk/myrevisionnotesdownloads**

Countdown to my exams

6–8 weeks to go

- Start by looking at the specification — make sure you know exactly what material you need to revise and the style of the examination. Use the revision planner on page 4 to familiarise yourself with the topics.
- Organise your notes, making sure you have covered everything on the specification. The revision planner will help you to group your notes into topics.
- Work out a realistic revision plan that will allow you time for relaxation. Set aside days and times for all the subjects that you need to study, and stick to your timetable.
- Set yourself sensible targets. Break your revision down into focused sessions of around 40 minutes, divided by breaks. These Revision Notes organise the basic facts into short, memorable sections to make revising easier.

REVISED ☐

2–6 weeks to go

- Read through the relevant sections of this book and refer to the exam tips, topic summaries, typical mistakes and key terms. Tick off the topics as you feel confident about them. Highlight those topics you find difficult and look at them again in detail.
- Test your understanding of each topic by working through the 'Now test yourself' questions in the book. Look up the answers at the back of the book.
- Make a note of any problem areas as you revise, and ask your teacher to go over these in class.
- Look at past papers. They are one of the best ways to revise and practise your exam skills. Write or prepare planned answers to the exam practice questions provided in this book. Check your answers online and try out the extra quick quizzes at **www.hoddereducation. co.uk/myrevisionnotesdownloads**
- Use the revision activities to try out different revision methods. For example, you can make notes using concept maps, spider diagrams or flash cards.
- Track your progress using the revision planner and give yourself a reward when you have achieved your target.

REVISED ☐

One week to go

- Try to fit in at least one more timed practice of an entire past paper and seek feedback from your teacher, comparing your work closely with the mark scheme.
- Check the revision planner to make sure you haven't missed out any topics. Brush up on any areas of difficulty by talking them over with a friend or getting help from your teacher.
- Attend any revision classes put on by your teacher. Remember, he or she is an expert at preparing people for examinations.

REVISED ☐

The day before the examination

- Flick through these Revision Notes for useful reminders, for example the exam tips, topic summaries, typical mistakes and definitions.
- Check the time and place of your examination.
- Make sure you have everything you need — extra pens and pencils, tissues, a watch, bottled water, sweets.
- Allow some time to relax and have an early night to ensure you are fresh and alert for the examinations.

REVISED ☐

My exams

Eduqas A-level and WJEC AS Paper 1

Date:..

Time:..

Location:..

Eduqas A-level and WJEC AS Paper 2

Date:..

Time:..

Location:..

Eduqas A-level Paper 3 and WJEC A2 Paper 1

Date:..

Time:..

Location :...

WJEC A2 Paper 2

Date:..

Time:..

Location:..

1 Changing landscapes

Coastal landscapes

The operation of the coast as a system

The coastal zone can be viewed as a system with:

- **inputs** — energy from wind, waves and tides; sediment from weathering and erosion processes
- **outputs** — sediment removed by longshore drift and sediments deposited as landforms, such as dunes
- **transfers** — processes of erosion and transportation that can move sediment around the system
- **stores** — sediment deposited in landforms

The coast is an open system. Energy and sediment can move from the boundary of the system into the environment around it.

Coastal sediment budget

The balance between the input, store and output of sediment is called the **sediment budget** (Figure 1.1).

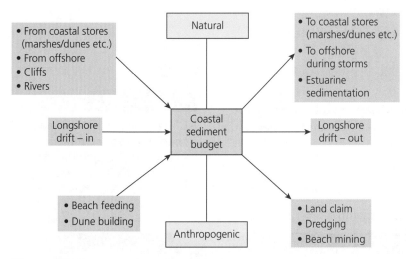

Figure 1.1 **The coastal sediment budget**

- When sediment input = sediment output, the budget is balanced (in a steady-state equilibrium).
- If factors such as human activity reduce the sediment input, the budget will be out of balance.
- When the system in one place is out of balance, it can increase erosion further along the coastline.

Revision activity

Outline how human activities could reduce or increase the supply of sediment to the sediment budget.

Sediment cells

- There are 11 major cells for England and Wales (Figure 1.2), divided into smaller sub-cells.
- Sediment can be lost from cells, making them open systems.

> A **sediment cell** is a stretch of coastline where the sediment budget is self-contained. Also known as a **littoral cell**.

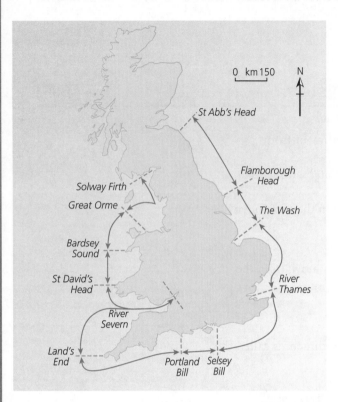

Figure 1.2 The coastal sediment cells in England and Wales

Dynamic equilibrium

Coasts are in a state of equilibrium when the inputs and outputs of the system are equal. Coasts are constantly changing due to variations in the energy conditions, such as storms.

There are three types of equilibrium (Table 1.1).

> **Typical mistake**
>
> Cliff erosion is not the main source of sediment input. Around 90% of sediment input comes from rivers. Only up to 5% comes from coastal erosion.

Table 1.1 The different types of coastal equilibrium

Type of equilibrium	Description	Example
Steady state equilibrium	Changes in energy and the resulting change in coastlines do not vary much from the long-term average conditions.	Beach profile adjusts in summer and winter as wave energy changes, but the average gradient stays the same.
Metastable equilibrium	The coastal zone changes from one state of equilibrium to another due to an event causing a change in conditions.	Sediment removal due to dredging changes the beach profile, or it disappears. There is a new equilibrium with a reduced beach.
Dynamic equilibrium	The state of equilibrium changes over a longer timescale than metastable equilibrium.	Climate change causing rising sea level allows new areas of land to be influenced by wave attack, resulting in a change in cliff profile.

Feedback

Feedback occurs as a result of a change in the coastal system, helping the system adjust.

- Positive feedback — increases the initial change that occurred.
- Negative feedback — reduces the effect of the change, helping the coast return to its original condition.

Now test yourself

TESTED

1 Why can the coast be classified as an open system?
2 What is a sediment cell?
3 Why are some coasts in a state of dynamic equilibrium?
4 What is the difference between positive and negative feedback?

Answers on p. 169

Temporal variations and their influence on coastal environments

REVISED

Tides, currents and waves all input energy into the coastal system.

Tides

- **Tides** created by the gravitational effect of the Moon (and to a lesser extent the Sun), which pulls water on Earth towards it to create high tides, with a balancing increase in sea level on the opposite side of the Earth.
- Between these two areas the sea level is lower, creating low tides.
- Movements of the Earth, Moon and Sun result in changes in the gravitational pull and changes in tides.
- Twice a month the Earth, Moon and Sun are aligned, so the gravitational pull is greatest, creating higher than average tides called **spring tides**.
- When the Sun and Moon are at right angles to each other the gravitational pull is weaker, creating lower than average tides called **neap tides**.

The **tidal range** is the difference in height between high and low water during the monthly tidal cycle. It influences the zone where coastal processes occur.

Exam tip

Make sure you understand the terms used in the specification, for example temporal variations are changes that occur over time, ranging from very short-term changes (lasting seconds) to much longer-term ones (over centuries and millennia).

Tides are the rise and fall of sea level. **Semi-diurnal tides** have two high tides and two low tides every 24 hours, while **diurnal tides** have one high tide and one low tide.

Offshore and onshore currents

Offshore and onshore currents are flows of water occurring in the coastal zone. There are three types:

- **Tidal currents** — water floods the intertidal zone at high tide, moving and depositing sediment. As the tide falls (**ebb tide**), sediment moves in the reverse direction.
- **Shore normal currents** — waves align parallel to the coastline, pushing water straight up the beach. Water returning directly away from the shore can form strong, fast-moving channels of water called **rip currents**.
- **Longshore currents** — waves approach the shoreline at an oblique angle but return straight down the beach, moving sediment parallel to the shore.

Exam tip

Ocean currents refer to the permanent, direct movement of water in the Earth's oceans, which can affect coastal environments by influencing the climate. Onshore/offshore currents are on a more local scale, and have a greater impact on coastal processes and landforms.

Constructive and destructive waves

Waves are created by the transfer of energy from the wind blowing over the sea surface. Stronger winds increase frictional drag and the size of waves.

Waves can be classified as constructive or destructive (Figure 1.3 and Table 1.2).

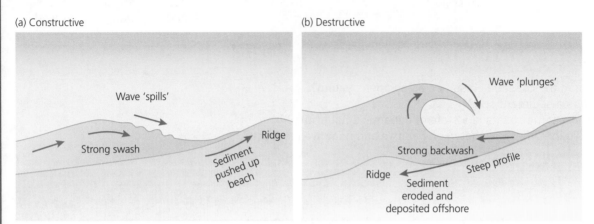

(a) Constructive

(b) Destructive

Figure 1.3 Constructive and destructive waves

Table 1.2 Features of constructive and destructive waves

Feature	Constructive waves	Destructive waves
Wave height	Less than 1 metre	Greater than 1 metre
Wave length	Long — up to 100 metres between crests	Shorter — 20 metres between crests
Wave period	Longer — 6–8 breaking each minute	Shorter — 10–14 breaking each minute
Wave steepness	Gentle — tend to spill over	Steep — plunging
Wave energy	Low	High
Stronger swash or backwash?	Swash	Backwash
Movement of material	Up beach to form a berm	Down beach
Beach gradient	Increase gradient of upper beach	Steeper upper beach (storm beach); gentler lower beach

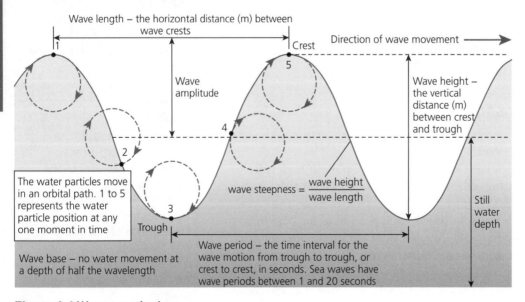

Figure 1.4 Wave terminology

Exam practice answers and quick quizzes at **www.hoddereducation.co.uk/myrevisionnotesdownloads**

Destructive waves are usually more frequent in winter, with stormier weather. This is due to the jet stream steering more Atlantic depressions over the UK during winter. Atlantic depressions occur more frequently and are more intense during winter months.

> **Typical mistake**
>
> Do not assume that destructive waves create erosional landforms, such as those associated with cliff retreat. Nor do they remove all the sediment or flatten the entire beach. The term refers to the fact that they move sediment down a beach, changing the gradient in different parts, and creating ridges and flatter sections. Constructive waves help to build up a beach.

Landforms and landscape systems — their distinctive features and distribution

REVISED

Coastlines can be divided into high-energy and low-energy coastal environments.

High-energy coastal environments

These are erosive, rocky coastlines. Processes affecting them include:
- physical, chemical and biological weathering
- mass movement, e.g. rock falls
- transportation processes moving material from or along the coastline

Examples of high-energy coastal landforms include cliffs and wave-cut platforms (p. 15).

Low-energy coastal environments

Deposition is the dominant process, creating sandy coastlines and associated features such as sand dunes, spits and bars. Estuarine coastlines are low-energy coastlines where features such as mudflats develop (p. 21).

> **Now test yourself**
>
> TESTED
>
> 5 What factors influence the type of waves found in a coastal area?
> 6 How do coastal landforms differ between high-energy and low-energy coastal zones?
>
> Answers on p. 169

Factors affecting coastal processes and landforms

REVISED

Waves

Wave type depends on:
- wind velocity — faster winds produce bigger waves
- length of time the wind has blown across the water — longer periods produce bigger waves
- **fetch** — the greater the fetch, the bigger the waves

> The **fetch** is the distance the wind has blown over the water.

The shape of the coastline and its orientation to oncoming waves affect the impact a wave has. A coastline may be at the receiving end of a long fetch, but its orientation may protect it from high-energy waves.

Wave refraction

Wave direction approaching the shoreline is modified due to the shape of the seabed (Figure 1.5).

The depth of water around a coastline varies. Friction from the seabed in shallow water slows the progress of waves.

↓

Waves change direction, so that they approach the coastline aligned parallel to it.

↓

This distorts the spread of energy — concentrated at headlands and dissipated in bays.

↓

Concentrated energy at headlands encourages erosion. Lower energy in bays results in sediments deposited, creating beaches.

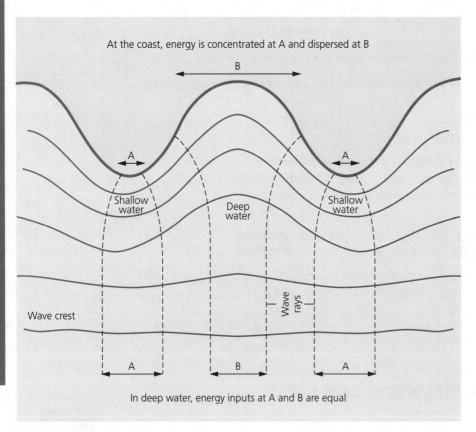

Figure 1.5 **Wave refraction**

Wave reflection

A wave hitting a vertical surface such as a cliff face or sea wall bounces back without breaking or losing its energy. The lines of incoming and outgoing waves are called **standing waves** or the **clapotis effect**.

Lithological factors

Rock type and **structure** have a significant effect on processes and the landforms created.

Hardness

- Harder, more resistant rock, such as granite, erodes slowly, often producing high cliffs.
- Rocks such as clays and sandstones have less structural strength and erode quicker. Slumping can be a dominant feature.

Composition

- Mineral composition affects the rate of weathering, for example sandstone made of silica is chemically inert, having a low rate of chemical weathering.
- Feldspars in granite can be changed by hydrolysis into clay minerals, decreasing resistance to erosion.
- Rocks such as chalk and limestone have a calcium carbonate content that is soluble in saltwater (p. 14).

Structure

- Rocks with many joints and bedding planes are weaker and more vulnerable to weathering and erosion.
- Faults are major lines of weakness where erosion processes can have even greater impact.
- Folding weakens rocks and affects the dip of rock strata, influencing the rate of erosion and cliff profiles.

> **Structure** refers to the joints, bedding planes, faults and folds in the geology of the area.

> **Exam tip**
>
> Rather than just naming geology as a factor, be specific about what aspect of the geology is important. Is it hardness, composition, joints, bedding planes, faulting, folding, dip, or a combination of factors?

Processes of coastal weathering, mass movement, erosion and associated landforms

REVISED

Marine erosion processes and sub–aerial processes (Table 1.3) combine to create distinctive landforms.

Table 1.3 Sub-aerial and marine erosion processes

Sub-aerial processes (cliff face processes)	Marine erosion processes (cliff foot processes)
Physical/mechanical weathering	Hydraulic action
Chemical weathering	Corrasion/abrasion
Biotic weathering	Attrition
Mass movement	Corrosion

Sub-aerial processes

Weathering

Table 1.4 Physical/mechanical weathering processes

Process	Description
Freeze–thaw	Repeated freezing and thawing of water results in the expansion of cracks in rocks, causing small fragments to break off
Salt crystallisation	Sea water evaporates from cracks, allowing salt crystals to grow, exerting pressure and causing pieces of rock to break off
Water-layer weathering	Constant wetting and drying (e.g. due to tides) causes clay-rich rock to expand and contract, resulting in cracks, which aid physical weathering processes

Table 1.5 Chemical weathering processes

Process	Description
Solution	The removal of rock dissolved in acidic rain water
Oxidation	Oxygen dissolved in water reacts with minerals, causing oxidation
Hydration	Minerals in rocks absorb water, which weakens their structure, making them susceptible to weathering
Hydrolysis	Breakdown of rock by acidic water (e.g. from **carbonation**) produces clay and soluble salts, especially feldspar in granite
Chelation	Organisms produce substances called chelates, which decompose minerals
Carbonation	Carbon dioxide dissolved in rain water creates weak carbonic acid, which dissolves calcium carbonate in limestone — this process plays an important role in the carbon cycle (p. 86)

Table 1.6 Biotic weathering processes

Process	Description
Plant growth	Roots of growing plants force their way into cracks and joints in rocks, increasing the pressure, and causing rock particles to fracture
Animals	Burrowing animals create weaknesses in soft rocks

Mass movement

Table 1.7 Mass movement processes

Speed	Process	Description
Rapid	Landslide	Rocks affected by physical weathering or marine erosion collapse downwards
	Rockslide	Rocks slide down a cliff face when the bedding planes dip towards the sea
Variable speed	Rotational slip	Softer rocks give way, moving downhill in one mass along a concave slip surface
	Slump/mudslide	Saturated soft rock (often on top of impermeable rock) flows downhill
Slow	Soil creep	Soil particles move downslope, aided by rain drop impact
	Solifluction	Movement of wet soil downslope caused when underlying layers are frozen

> **Weathering** is the disintegration and breakdown of rocks in situ by the action of the weather, plants and animals.

> **Mass movement** is the downward movement of material due to gravity. It varies in speed from sudden and rapid to very slow.

Marine erosion processes

Five main types of marine erosion process can be active on a stretch of coastline at any one time (Table 1.8).

Table 1.8 Marine erosion processes

Process	Description
Hydraulic action	Breaking waves create hydraulic pressure in joints. Air in cracks in a cliff face can be compressed by the force of the waves and then rushes out when the wave retreats. This can weaken the rock.
Wave quarrying	High-energy waves can exert a force of many tonnes per metre squared, which can remove loose, unconsolidated rock fragments.
Abrasion/corrasion	Rock fragments wear away the coast. Abrasion involves rock particles being scraped over bare rock, wearing it away and smoothing it. Corrasion occurs when waves hurl debris against the rock, causing pieces to be broken off and wearing it away.
Attrition	Eroded rocks are worn smaller and rounder by constant rubbing against each other with the movement of the sea.
Corrosion/solution	Weak acidic sea water chemically attacks certain rocks, dissolving minerals.

Characteristics and formation of coastal erosional landforms

Distinctive coastal landforms can be large scale, such as cliffs, wave-cut platforms, headlands and bays, or small scale, such as caves, arches, stacks, blowholes and geos.

Cliffs

A cliff is a steep slope that is affected by marine processes. The cliff profile is influenced by the following factors:
- **Geology** — harder rocks usually produce steeper cliffs, while softer rocks form lower-angled cliffs.
- **Structure** — the dip of the bedding planes influences the angle of the cliff.

Cliff retreat and the formation of wave-cut platforms

Marine erosion creates a wave-cut notch at the cliff base.

The cliff is undercut, collapses and the process repeats.

The cliff retreats, leaving a gently sloping surface (1–5°) called a **wave-cut platform**, also called an **intertidal shore platform**.

A wide wave-cut platform prevents marine erosion processes reaching the cliff foot except in storms or the highest tides, reducing the rate of erosion.

Sub-aerial processes continue working on the cliff face, reducing the angle of slope (Figure 1.6).

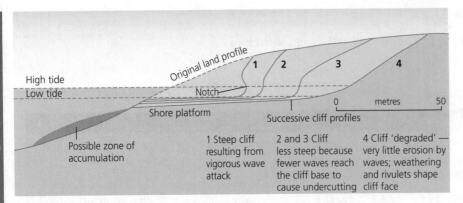

Figure 1.6 Cliff retreat and the formation of a wave-cut platform

Headlands and bays

Differences in geology result in rocks eroding at different rates, especially on **discordant coastlines** (Figure 1.7).

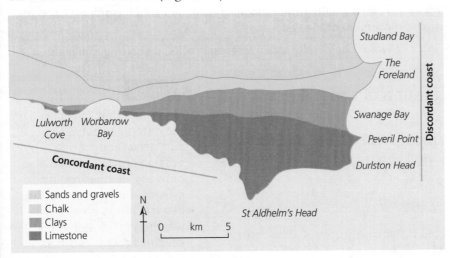

Figure 1.7 The role of geology in the formation of headlands and bays on the Isle of Purbeck coast

Cave, arch, stack, stump sequence

Marine erosion processes create **sea caves** along a line of weakness.

Further erosion results in caves either side of a small headland meeting, forming an **arch**.

Over time the arch becomes unstable and collapses, leaving an isolated pillar of rock — a **stack**.

Erosion processes attack the stack causing it to collapse, leaving a short section — a **stump**.

On a **discordant coastline** bands of different rock types lie at right-angles to the shoreline (on a **concordant coast** beds of different rock types lie parallel to the shoreline).

Revision activity

Explain the factors that have influenced the formation of headlands and bays in the Isle of Purbeck area.

Typical mistake

Not all stacks form from an arch. Some may result from cliff collapse without an arch being formed.

Cave, blowhole, geo sequence

Marine erosion processes create **sea caves** along a line of weakness.

↓

Waves force air and water into the cave, creating a shaft upwards to the surface and forming a **blowhole**, where water gushes out in certain tide and wave conditions.

↓

Continued erosion causes the extended cave roof to collapse, leaving a long, narrow inlet — a **geo**. Geos can also form along a line of weakness, where erosion processes have a greater impact.

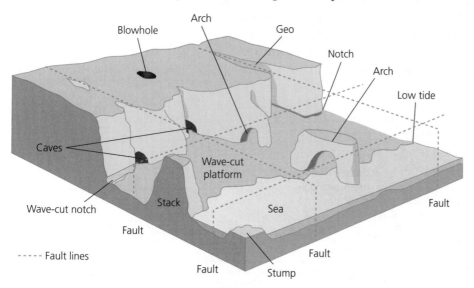

Figure 1.8 Landforms associated with coastal erosion

Now test yourself

TESTED ☐

7 What factors can affect the type of coastal processes and landforms found at the coast?
8 What are sub-aerial processes?
9 What are the processes involved in the formation of a stack?

Answers on p. 169

Processes of coastal transport and deposition and the associated landforms

REVISED ☐

Coastal transport

Sediment size and the strength of the current are key factors determining which method of transport occurs. There are four main methods of transport (Table 1.9).

Table 1.9 Methods of sediment transport

Method of transport	Description
Traction	Sediment rolls or slides along the seabed
Saltation	Sediment is bounced along the seabed
Suspension	Small particles are held and moved in the water
Solution	Material dissolves in the water and is moved by the sea

Longshore drift

Longshore drift is the process that moves sediment along the shoreline (Figure 1.9).

Waves approach at an angle due to prevailing winds.
Swash washes sediment diagonally up the beach.

↓

Gravity pulls backwash with sediment straight down the beach.

↓

Repeated action results in the net movement of sediment along the beach parallel to the shoreline. The direction is predominantly one way due to the prevailing winds. Changes in wind direction can result in movement in the opposite direction, bringing greater complexity to the landscape.

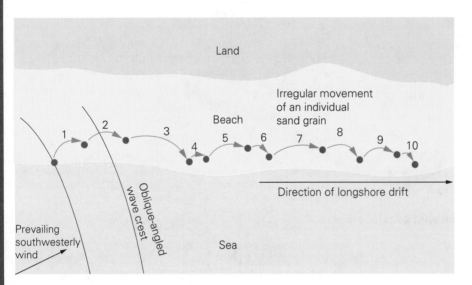

Figure 1.9 The process of longshore drift

Coastal deposition

Deposition occurs when the energy of the sea is too low to transport material. This occurs:

- when waves lose velocity after breaking
- when backwash percolates quickly into the beach
- when sheltered coastlines reduce the wave energy
- as a result of **flocculation**

Sediment sorting

- Large sediment is deposited furthest from the sea by high-energy waves, such as in a storm, creating a steep beach of larger material — a storm beach.
- Smaller material is deposited nearer the sea.
- Material nearer the sea is more prone to attrition and being eroded into smaller particles.

Flocculation occurs where fresh water mixes with sea water, such as in a river estuary. Clay particles coagulate due to chemical reactions to form flocs, which are heavier and more likely to be deposited.

Characteristics and formation of coastal depositional landforms

Beaches

- Beaches comprise loose, unconsolidated sand, pebbles and cobbles.
- The main influences on beach construction are sediment supply, wave energy and wave direction.
- Beaches are dynamic landforms, where the profile changes quickly according to changing wave energy.

There are three main types of beach (Table 1.10).

Table 1.10 Types of beach

Type of beach	Dominant waves	Sediment movement
Swash-aligned beach	Waves break parallel to the shore	Sediment moves onshore/offshore
Drift-aligned beach	Waves break at an oblique angle to the shore	Longshore drift along the shore
Zeta-formed beach	Waves break at an oblique angle but a headland at each end causes wave refraction, blocking sediment movement	Longshore drift along the coast, but sediment builds up in front of a headland

Spits

A spit is a linear ridge of sediment joined to the land at one end (Figure 1.10).

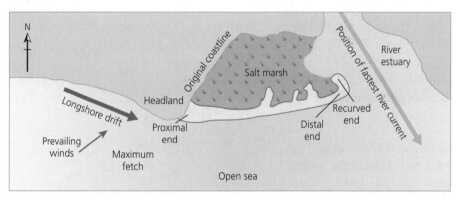

Figure 1.10 The features of a spit

- Longshore drift moves sediment along the coast.
- Where the coastline suddenly changes direction, the material continues to be deposited in the original direction.
- Short-term changes in wind direction or currents cause the **distal end** to curve, e.g. Hurst Spit in Hampshire.
- In the sheltered water behind the spit sediment is deposited to form mudflats or saltmarsh.
- Double spits occur where a pair of spits grow towards each other from facing sides of a bay, for example Christchurch Harbour in Dorset. (However, this example may have been formed by the sea breaching a bar formed across the bay.)

> The **distal end** of a spit lies in open water (the **proximal end** is joined to the mainland).

Bar

- A bar forms when a spit grows to form a ridge that blocks off a bay, creating a lagoon between the mainland and the bar. An example of a bar is Slapton Sands in Devon.

> **Exam tip**
>
> Make sure you know named and located examples of the different landforms.

- Alternatively, a bar may start as deposits out at sea that are driven onshore by rising sea levels (such as at the end of the Ice Age). As sea levels stabilise it becomes a relict feature, for example Chesil Beach in Dorset.
- If the bar is not connected to the mainland it is a barrier island, for example Scolt Head Island off the north Norfolk coast.

Tombolo

A tombolo is a spit or bar that joins the mainland to an island, for example the eastern end of Chesil Beach, where it joins the Isle of Portland.

Cuspate foreland

A cuspate foreland is a triangular-shaped projection from the shoreline made up of a series of ridges created by longshore drift from opposing directions. A famous example is Dungeness in Kent.

Aeolian, fluvial and biotic processes and the associated landforms in coastal environments

REVISED

Aeolian processes

When wind reaches a critical velocity it transports sediment particles, which can lead to the development of sand dunes. The conditions required are:
- an abundant supply of sand
- a shallow beach gradient
- a tidal range allowing a large area of sand to dry out at low tide
- prevailing onshore winds
- an area inland where windblown sand can accumulate

Formation of sand dunes

Constructive waves deposit sediment on the beach.

Winds blowing onshore move sand inland by:
- creep — sand is rolled along the surface
- saltation — smaller sand particles bounce over the surface

Where an obstacle or vegetation reduces wind velocity, the sand is deposited and quickly accumulates.

Plants such as sea couch grass and marram grass become established, increasing sand accumulation and forming an embryo dune.

Continued deposition leads to the embryo dunes forming a ridge.

New embryo dunes form in front of the ridge, repeating the process.

Gradually the older dunes become covered in vegetation and are stabilised (Figure 1.11).

> **Revision activity**
>
> Look at examples of coastlines shown on OS maps to get used to recognising the way each landform is depicted. If the landform is named, an internet search of images will allow you to view what it looks like.

> **Exam tip**
>
> An OS map extract can be used to locate coastal landforms. Make sure you can quote six-figure grid references accurately.

> **Aeolian** processes relate to wind activity and how the wind erodes, transports and deposits sediments.

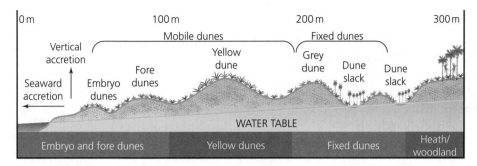

Figure 1.11 The features of a sand-dune landform

The sand dunes at Studland in Dorset contain all the features in Table 1.11. They started forming 500 years ago and are growing at 1 metre per year.

Table 1.11 The features of sand dunes

Sand dune features	Description
Ridges	Lines of dunes parallel to the coast
Slacks	Hollows found between the dune ridges
Grey dunes	Older dunes where decomposing organic matter creates a humus layer on the surface, making the dunes appear grey
Yellow dunes	Younger dunes with no humus layer
Embryo dunes	Youngest dunes in the early stages of formation
Fore dunes	Dune ridges closest to the sea
Fixed dunes	Older dunes stabilised with vegetation, which are unlikely to change
Blow-out	An area of dune that has been eroded by the wind, often due to its protective cover of vegetation being removed by animal or human activity

Fluvial processes

Fluvial processes of erosion and transportation provide a significant supply of sediment to the coastal zone. River estuaries are low-energy coastal environments, with deposition the dominant process.

Tidal flats

Tidal flats are level, muddy areas bordering an estuary.
- Edges of estuaries are sheltered and so are low-energy coastlines.
- Deposition of fine sediment is the main process creating a tidal mudflat.
- At low tide the mudflat is uncovered. Fresh water from tributaries flows across the mudflat to the sea in small channels and **rills**.
- Morecambe Bay on the Lancashire coast is an example of tidal flats.

> **Revision activity**
>
> A blow-out resulting from a loss of sand dune vegetation leads to higher wind velocities and more sand being transported, which makes it harder for vegetation to re-establish.
>
> How is this an example of positive feedback? What could be done to reduce the impact of a blow-out?

> **Rills** are shallow channels cut by the action of running water.

Salt marshes

- Salt marshes are gently sloping, vegetated areas of intertidal mudflats.
- Salt marshes are found in low-energy, sheltered locations such as estuaries, lagoons and behind spits.
- A lack of marine erosion allows salt-tolerant, pioneer plant communities to colonise the mudflats.
- Salt marshes are dissected by channels and rills.
- They are important wildlife habitats, some having been protected and managed for wildlife.
- Many areas in the past have been drained and used for agriculture, for example the Blyth estuary in Suffolk.
- Many examples of salt marsh are found along the 120 kilometres of Essex coast between the Blackwater and Crouch estuaries. It is known as 'the Saltmarsh Coast'.

Biotic processes

Coral reefs

Coral reefs are calcium carbonate structures secreted by tiny animals called polyps. They build slowly over time and can be hundreds of metres thick.

The best conditions for coral growth are:
- clear water, allowing light penetration
- seawater temperatures 23–29°C
- shallow water — less than 100 metres deep
- wave activity, which aerates the water

91% of all coral reefs are found in the Indo-Pacific region. The Great Barrier Reef off the Queensland, Australia coast is the largest, with over 2,900 individual reefs.

Changes in conditions can kill coral, making it susceptible to erosion. Broken coral can be transported and deposited to make rubble mounds suitable for new coral colonisation.

> **Exam tip**
>
> The study of coral reefs is an example of synoptic links between different components of the specification. As well as the study of their formation in Coastal landscapes, attempts to protect and manage an area of coral reef can be studied in Global governance (p. 116).

Mangroves

- Mangroves form an ecosystem of small trees growing along tropical coastlines where fine sediments have been deposited.
- The vegetation is adapted to be salt tolerant and able to survive in waterlogged mud with extremes of temperature.
- Adaptations include stilt roots, roots that grow upwards from the mud to allow oxygen uptake and the ability to absorb air through the bark.
- Mangroves encourage further deposition and protect coastlines from erosion by reducing wave energy. This can be a method of reducing the impacts of tsunamis (p. 160).
- The Great Sundarbans mangrove forest in India and Bangladesh is the largest in the world, covering 1,400 km².

Now test yourself

TESTED ☐

10 What may happen to depositional features along a coastline if human activity reduces the supply of sediment by longshore drift?

11 What role does vegetation play in the formation of a sand dune environment?

Answers on p. 169

Variations in coastal processes, landforms and landscapes over different timescales

Coastal processes vary over timescales ranging from seconds to millennia, resulting in changes to landforms and landscapes (Table 1.12).

Table 1.12 The timescale of coastal changes

Timescale		Example of change
Fast ↑ ↓ Slow	Seconds	High-energy storm activity
		Rapid mass movement
	Seasonal	Changes in beach profile
	Millennia	Sea-level change

Changes in seconds

High-energy storm events

A storm event or tsunami (p. 152) causes an increase in wave energy, resulting in increases in erosion and transportation, which can cause:
- removal of large amounts of beach sediment, changing the profile or removing it altogether
- destruction or breaching of sand dunes
- coastal flooding

Rapid mass movement

Sudden rockfalls, landslides and slumps create changes in cliff-face profiles and retreating cliffs, with the loss of land and possibly buildings.

Seasonal changes

Changes to the beach profile

Seasonal weather changes influence wave type, altering the beach profile (Table 1.13).

Table 1.13 Seasonal changes influencing a beach profile

Summer	Winter
• Fewer storms • Less frequent high wind speeds • Lower-energy waves • Waves predominantly constructive • Sediment moved onshore, building up the beach, increasing the gradient of the upper beach and forming a berm	• More storms • High winds more frequent • Higher-energy waves • Destructive waves more frequent • Sediment moved offshore, lowering the beach profile, creating a steeper upper beach and a gentler lower beach

Changes over millennia

Over geological time **sea levels** have changed by many metres. There are two types of change (Table 1.14).

> **Sea level** is the relative position of the sea surface as it comes into contact with the land.

Table 1.14 Types of sea level change

Eustatic change	Isostatic change
• A global change in the volume of water in the oceans • During a glaciation period more water is frozen, resulting in less liquid in the oceans, so sea level falls (p. 78) • Global warming of the climate increases melting of continental ice sheets, adding water to the oceans and raising sea level • Warming of the oceans results in the volume of water expanding, creating a rise in sea level	• A localised change in the relative sea level caused by the upward or downward movement of land masses • During glacial periods the weight of ice causes the land to sink into the crust, making sea levels appear relatively higher • Melting ice removes the weight and the land very slowly rises (**isostatic recovery**), causing a relative fall in sea level

> **Exam tip**
>
> For the WJEC/Eduqas specification you only need to learn *either* eustatic *or* isostatic changes in sea level. However, a knowledge of both is useful for understanding the formation of some coastal landforms.

The impact of changing sea levels on landforms

Rising sea levels

- Rising sea levels flood lower-lying parts of the coast.
- Deltas, spits and beaches disappear under water or due to increased rates of erosion.
- River floodplains and valleys flood to form a broad river estuary called a **ria**, for example the Kingsbridge estuary in Devon.
- If the flooded valley is a glaciated U-shaped valley, a **fjord** (or fiord) is formed.
- The shape of the valley means fjords are very deep, flat bottomed and steep sided, for example Sognefjord in Norway, which is 204 km long and 1,308 m deep.
- Fjords often have a shallower entrance called a threshold, formed as the glacier had less erosive power where the ice met the sea, or when the glacier deposited a terminal moraine. If the sea level rise does not cover the threshold a small, rocky island, called a **skerry**, may be created. Many of the islands off the coast of Norway are skerries, linked to the 1,190 fjords found along the coast.

Falling sea levels

- Beaches are no longer affected by waves and are left stranded above the new sea level, forming **raised beaches**.
- The former cliff line and landforms are left stranded as relict cliffs. These are no longer undercut and over time become degraded cliffs, due to sub-aerial processes, and covered in vegetation.
- The wave-cut platform appears raised above the new sea level to form a marine terrace, which can be used for agriculture.
- These features can be found along the western coast of Scotland, where the rate of isostatic recovery is around 2 mm a year.

> **Exam tip**
>
> For the WJEC/Eduqas specification you only need to learn the impact of sea level change on *one* landform.

Coastal processes are a vital context for human activity

REVISED

Positive impacts of coastal processes on human activity

Coastal zones are important for human activities, including the following:

- Recreation and tourism — coastal landscapes provide the opportunity for a wide range of leisure pursuits, and coastal landscapes and features attract people to the coastal zone.
- Residential — people want to live by attractive coastal landscapes.
- Job opportunities from tourism and industry attract people to the area. The EU estimates that coastal tourism creates job opportunities for 3.2 million people, generating €183 billion.
- Agriculture — tidal mudflats can be drained and used for farming, for example in parts of the river estuaries in Suffolk, such as the River Blythe, defences were built to prevent flooding and the land was drained.
- Industry — rias provide deep water ports needed for importing materials, for example Milford Haven in Pembrokeshire, Wales has a depth of over 17 m even at low tide.
- Transportation — rias allow deep water vessels to travel inland, for example the port of Southampton is 16 km inland at the head of Southampton Water. Flat marine terraces above sea level are ideal for road and rail communications, for example the railways and A8 and M8 roads along the River Clyde estuary west of Glasgow.

The growth of tourism

Tourists are attracted to the coastal zone for many reasons:

- Natural features:
  - attractive and dramatic scenery, for example the Pembrokeshire Coast National Park
  - sandy beaches, as found along the Spanish Costas
  - safe, warm seas with a lack of strong currents, for example around the Maldive Islands
  - large waves suitable for surfing, such as those found off the Gold Coast of Australia
  - ecosystems and their wildlife, for example the Galapagos Islands or the Great Barrier Reef off Australia, which is popular for snorkelling
  - fossils exposed by retreating cliffs, for example the Jurassic Coast in Dorset
- Deep water ports that are suitable for large cruise ships. Vancouver's cruise terminal can handle four cruise ships at a time and caters for around 900,000 passengers a year, contributing $2 million to the local economy.
- Rebranding and marketing by coastal resorts. Brighton, Southend-on-Sea and Hartlepool have all rebranded to attract tourists. Blackpool has rebranded to try to shed its downmarket image in favour of a more sophisticated and glamorous feel, coinciding with redevelopment of the town.
- To visit places seen in TV shows or films. For example, the TV series *Broadchurch*, first shown in 2013, was filmed around West Bay in Dorset and was watched by up to 10 million people per episode. In

2014, 77% of local businesses reported increased customer numbers with almost half thinking it was due to the programme. 70% of businesses reported increases in turnover.

In many coastal areas, tourism represents the most important economic activity. In 2016, seaside tourism was worth £8 billion in England.

Tourism at the coastal zone can have a number of impacts:
- The development of services and infrastructure to accommodate tourists. In the 1960s, the plan to encourage tourism resulted in Benidorm growing rapidly from a small fishing village into a town of 69,000 people today, catering for over 4 million tourists a year.
- Negative impacts on the environment, such as:
  - footpath erosion. Over 200,000 people per year walk the coast path between Lulworth Cove and Durdle Door in Dorset. Over time the path has widened by 2 m and eroded downwards by 30 cm, resulting in the whole of the adjacent hillside beginning to slip.
  - damage to ecosystems. In Cancún in Mexico, 57 hectares of mangrove forest were cleared in 2016 for the construction of a tourist complex
  - unsustainable demand for water. The Spanish Costa del Sol has over 60 golf courses, and golf tourists spend more than any other group. A golf course can use 700,000 m^3 of water a year, which is enough for a town of 25,000 people. Combined with the demand from new hotels catering for the tourists, this drought-prone area is unlikely to be able to meet future demand
- Infrastructure overload, such as traffic congestion.
- An increase in second homes in attractive holiday locations. In Cornwall, the towns of St Ives, Fowey and Mevagissey have all voted to ban second home owners buying new developments.
- Facilities only cater for tourists and are closed out of season.
- Socio-cultural changes, such as:
  - commercialisation of local culture. Local customs and traditions may be altered to conform to tourist expectations — known as reconstructed ethnicity
  - locals may resent the influx of tourists and their cultural ideas. In Benidorm, many locals resented the 'lager lout' image of its visitors in the past and the town had to work hard to attract different tourists — an example of rebranding (p. 72)
- Vulnerability of reliance on one economic activity, for example a terrorist attack results in tourism declining overnight. The terrorist attack in Tunisia in 2015 resulted in a 25% drop in the number of visitors, mass unemployment and a $1.1 billion decrease in revenue the following year.
- Cruise ship tourists spend little in the coastal area because they have no need for the services. Cruise ships can be a source of pollution at sea (p. 129).
- A large number of jobs are created, although many are poorly paid and seasonal. Higher-paid managerial jobs may go to foreigners, such as hotel managers in a global hotel chain.
- Local people make a living by providing services for tourists.
- Money earned by locals is spent locally, creating a multiplier effect.

Revision activity

Classify the impacts of the growth of tourism into environmental, economic and social.

Negative impacts of coastal processes on human activity

Coastal processes, especially erosion, have a number of negative impacts on human activity:

- Rapid mass movement events, such as cliff collapse, are hazardous to life but kill very few people in the UK.
- Coastal erosion, resulting in cliff retreat, causes damage or loss of buildings and infrastructure, and the loss of valuable agricultural land and therefore income for farmers. The British Geological Survey estimates that 113,000 residential and 9,000 commercial properties, as well as 5,000 hectares of farmland, are at risk, with a value of £7.7 billion. Rising sea levels resulting from climate change will increase the level of vulnerability for many areas (p. 95).
- A loss of beach sediment results in a decline in the tourism industry and a loss of income.
- The erosion of vulnerable ecosystems, such as coral reefs, results in a loss of tourism.
- Sea-level rise makes low-lying coastal areas vulnerable to flooding. Coastal cities in Brazil and the Pacific Islands are vulnerable. Worldwide, damage could cost $1 trillion by 2050.
- An increase in sediment input results in increased deposition further along the coast, affecting harbours and ports.

Management strategies relating to impacts of coastal processes on human activity

There are five strategies that can be adopted (Table 1.15).

Table 1.15 Coastal management strategies

Management strategy	Description
Do nothing	Allows natural processes such as coastal erosion to continue
Managed retreat or realignment	Allows the shoreline to move inland by erosion or flooding to a new line of defence
Hold the line	The present shoreline is protected by a variety of hard and soft engineering solutions
Advance the line	The shoreline is moved seawards either using hard engineering structures or by encouraging sand dune growth
Limited intervention	Deals with the problem to some extent, for example by encouraging the growth of salt marsh or sand dunes that reduce the impact of wave energy; it can also involve raising buildings to cope with flooding due to rising sea levels

> **Exam tip**
>
> Make sure you have detailed, up-to-date knowledge of one located management strategy. You also need an understanding of the other potential strategies.

The choice of management strategy can depend on the following criteria:

- **Feasibility** — considers the technical aspects, for example is an engineering solution possible given the marine processes and factors such as geology?
- **Cost–benefit analysis (CBA)** — divides the value of benefits (e.g. property protection and employment) by costs (e.g. capital building

costs and maintenance). Benefits should outweigh costs for the adoption of a strategy.

- **Environmental impact analysis (EIA)** — considers the effects on the environment, such as changes to sediment flows, on the immediate area and further along the coast.
- **Risk assessment** — considers factors such as the recurrence intervals of storms, what the strategy is designed to prevent and how long it should last. It weighs these up against the value of what is at risk, such as people and property.

A coastal management strategy for erosion may use one, or a combination, of the methods listed in Table 1.16.

Table 1.16 Methods of protecting against coastal erosion

Hard engineering	Soft engineering	Managed retreat
Control erosion and protect areas. For example: ● **Sea walls:** concrete structures at the cliff foot to absorb wave energy ● **Rock armour (rip-rap):** large rocks placed to absorb wave energy ● **Revetments:** wooden or concrete structures that absorb wave energy while allowing some flow of sediment ● **Gabion cages:** steel cages filled with small rocks to add strength to a coastline ● **Groynes:** wooden structures jutting into the sea to trap longshore drift ● **Drainage:** drains in cliffs to remove water and thus prevent land slips ● **Offshore bars:** islands of boulders offshore to absorb the force of the waves before they reach land ● **Rock bund:** a row of rocks along a beach	Work with natural processes; usually based on preserving the beach. For example: ● **Beach nourishment (replenishment):** sand or shingle is added, extending the beach or replacing eroded material ● **Beach reprofiling:** beach shape is changed to absorb more energy, reducing erosion ● **Beach recycling:** sediment is moved along the beach to counteract longshore drift ● **Fencing/hedging:** preserves beaches or dunes by reducing the amount of sand blown away ● **Replanting vegetation:** planting grasses or salt-resistant plants to stabilise areas, reducing erosion ● **Cliff profiling:** reducing the cliff angle, making the cliffs more stable	Also called coastal realignment. Existing coastal defences are abandoned, allowing the sea to flood inland until it reaches higher land or a new line of defence. It may allow the development of salt marshes.

Case study: Strategy to manage the impacts of coastal processes on human activity

Walton-on-the-Naze, northeast Essex coast

- Coastline consists of cliffs of London clay with various sand and gravel beds on top.
- The cliffs rise 23 metres, gradually dropping to the level of the beach 300 metres further north.
- The cliffs retreat by slumping and rotational shear, with cliff foot erosion removing the debris. The average rate of retreat between 1954 and 2000 was 2 m per year.

- Housing exists on the southern part of the cliffs.
- At the highest point is the Naze Tower, a 26 m high navigational aid built in 1720 and now Grade II★ listed.
- North of the tower is an area of public open space.
- The area is a Site of Special Scientific Interest (SSSI) due to its geology.

Table 1.17 Management strategy for Walton-on-the-Naze

	High cliffs south of the Naze Tower	High cliffs close to the Naze Tower	Northern cliffs
Management policy until 2105	Hold the line by maintaining existing defences	Managed retreat	No active intervention
Strategy	Cliff profiling to reduce angle of slope Sea wall Groynes Scheme established 1977–1980 at a cost of £1 million Houses have been protected and cliffs are stable	16,000 tonnes of granite rock armour placed along 110 m of the cliff foot nearest the Naze Tower 'Crag Walk' allows visitors to walk along the defence even at high tide Built in 2011 at a cost of £1.2 million Allows the cliff to collapse but prevents cliff foot erosion, so the cliffs eventually form a stable angle; cliff top should be stable before it reaches the Tower Life expectancy of defences is 50 years	Shoreline allowed to develop naturally

The impact of human activity on coastal landscape systems

REVISED

Positive impacts of human activity

Human activity has a positive impact on coastal processes and landforms, which is seen in present-day management and conservation strategies. These include:

- adoption of sustainable management and conservation of coastal environments, using soft engineering techniques
- increasing use of integrated shoreline management strategies, which consider all the conflicting needs and constraints along the coast rather than taking a piecemeal approach

Conservation strategies

Conservation ranges from total protection through to strategies allowing varying degrees of human activity. They vary in scale from very large to small areas of the coast. Examples include:

- World Heritage sites (an example of global governance — see also p. 133), for example the Great Barrier Reef in Australia and the Jurassic Coast in Dorset
- National Marine Reserves, called Marine Conservation Zones in England and Wales, for example Skerries Bank and Surrounds in Devon
- Sites of Special Scientific Interest (SSSIs), for example Sefton Coast, including the Ainsdale Sand Dunes National Nature Reserve in Merseyside

Negative impacts of human activity

The coastal zone is the area where natural processes and human activities come into greatest conflict, especially as demand for human use of coastal areas increases. This demand can be for a variety of economic uses, for recreation, or for conservation.

Offshore dredging

Dredging is the extraction of sand and gravel from the seabed for use in construction, especially of sea defences. It provides the material for beach nourishment schemes. Material is removed to improve shipping access.

Negative effects include:

- destruction of seabed habitats and the marine food web
- changes to wave types and sediment flows, resulting in changes in the beach profile, which can increase the rate of erosion if waves travel further up the beach

> **Typical mistake**
>
> Do not just consider the negative aspects of human activity. Remember that conservation and management can have a positive impact on the coastal zone.

Erosion of sand dunes

Sand dunes are fragile environments, which are vulnerable to damage from human activity. There are four types of impact (Table 1.18).

Table 1.18 Human impacts on a sand dune environment

Impact	Description
Conversion	The dune area is used for urbanisation or activities such as golf courses. Natural vegetation is altered or removed, changing the dune environment.
Removal	Sand is removed for another use, such as industry, or to improve transport or access to the beach. This allows further erosion by the wind.
Overuse	Dunes are used as an amenity for recreation activities, tourism or as military training areas. Overuse results in vegetation removal, increasing the rate of wind erosion.
External factors	Human activity in another part of the coast affects the dune system. For example, the building of sea defences reduces sediment input to the dunes.

Impacts can be reduced by management, for example:

- complete reconstruction in heavily damaged areas
- restoration by revegetation and fencing to limit access
- removal of external factors

Managing the impacts of human activity on coastal processes and landforms

Most countries now adopt the preferred strategy of **integrated coastal zone management (ICZM)**:

- ICZM is a form of sustainable management, which tries to balance environmental, economic, social, cultural and recreational needs within the limits controlled by natural factors.
- It involves the participation of all stakeholders in the coastal zone, and aims to meet the needs of all involved.
- Where applicable, it uses defence methods that enhance the environment.
- It aims to meet the needs of the present population while maintaining the potential of the area for future generations.
- Increasingly, management has to work across boundaries, both within nations and internationally.

Case study: Strategy to manage the impacts of human activity on coastal processes and landforms and landscapes

Ainsdale sand dunes, south Lancashire

- Ainsdale sand dunes cover $7\,km^2$ of the Sefton Coast sand dune system.
- They are recognised as an area of special interest for wildlife.
- Large numbers of visitors are attracted to the area. 5 million people live within 1 hour's drive of the dunes.
- People are attracted by the wide beaches, but also enjoy walking in the dunes, and driving quad bikes and 4×4 vehicles. Fires are lit for barbecues.
- Footpath erosion and other activities remove the vegetation, which can result in blow-outs and rapid erosion of the dunes by wind.
- The Ainsdale dunes are a National Nature Reserve managed by Natural England.

- The management strategy includes the following:
  - A zoning system is used on the Sefton Coast as a whole, which recognises honeypot areas such as Ainsdale Beach.
  - In honeypot areas intensive management occurs, providing facilities for visitors, such as car parking.
  - Where visitors are allowed, footpaths are covered in boardwalks to prevent erosion.
  - Attempts have been made to prevent vehicles entering the dune system.
  - Access to the nature reserve itself is by permit only, limiting numbers and so protecting the wildlife habitat.
  - Wardens are employed to manage the area.
  - Restoration work, such as planting in damaged areas, is carried out.
- Since 2010 there has been a reduction in the amount of money spent on management. Fewer wardens are employed to patrol the area, and there are fewer repairs to fences and boardwalks. As a result, more vehicles are driving into the dunes, leading to more damage.

Now test yourself

TESTED

12 What is:
 a) eustatic sea-level change?
 b) isostatic sea-level change?
13 Why is it important to take coastal sediment cells into consideration when deciding on a coastal management strategy?

Answers on p. 169

Glaciated landscapes

The operation of a glacier as a system

REVISED

Glaciers can be viewed as open systems with:
- **inputs** — energy from the Sun, precipitation, rock debris
- **outputs** — meltwater, sediment deposition
- **transfers** — erosion and transportation processes, which move ice and rock around the system
- **stores** — the **glacier**

Energy, water and sediment move from the boundary of the system into the environment around it.

The glacial budget

Dynamic equilibrium

Glaciers are in a state of equilibrium when the inputs and outputs of the system are equal (Figure 1.12). They are constantly changing due to variations in parts of the system.

> A **glacier** is a large body of ice formed from compressed snow. There are several types (see p. 38, Table 1.22). Glaciers are part of the cryosphere (p. 78).
>
> The **glacial budget** is the balance between inputs to and outputs from a glacier.

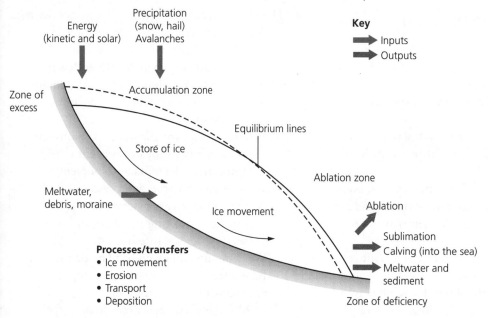

Figure 1.12 The glacier mass balance

There are three types of equilibrium (Table 1.19).

Table 1.19 Three types of glacier equilibrium

Type of equilibrium	Description	Example
Steady-state equilibrium	Changes in accumulation and ablation do not vary much from the long-term average conditions.	The glacier adjusts in winter and summer due to changes in temperature, but the average size stays the same.
Metastable equilibrium	The glacier changes from one state of equilibrium to another due to an event causing a change in conditions.	Subglacial volcanic activity increases melting of the ice. When the activity ends there is a new equilibrium with a reduced glacier.
Dynamic equilibrium	The state of equilibrium changes over a longer timescale than metastable equilibrium.	Climate change causing increased temperatures results in ablation being continually greater than accumulation. The glacier reduces in mass, leaving areas of land uncovered.

- When accumulation (input) is greater than ablation (output) the glacier grows in mass, usually in the upper section of the glacier.
- When accumulation is less than ablation the glacier shrinks in mass, usually in the lower section.
- The part of a glacier where accumulation and ablation are in balance is known as the equilibrium point.

Short-term changes

- In winter's colder conditions accumulation usually exceeds ablation, giving a positive balance.
- In warmer weather ablation usually exceeds accumulation, giving a negative balance (Figure 1.13).

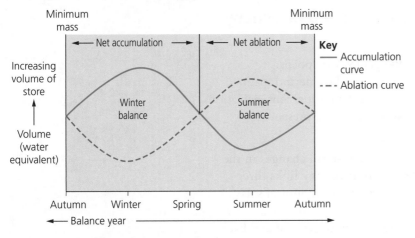

Figure 1.13 The annual mass balance of a glacier

Long-term changes

- Glaciers respond to long-term changes, such as changes in accumulation due to climate change.
- There have been a number of **glacial** and **interglacial** periods during Earth's history.

Feedback

Feedback occurs as a result of a change in the glacier system, helping the system adjust.

- Positive feedback increases the initial change, for example:

Increase in snow cover

More solar energy reflected, creating cooler temperatures

More snow and ice build up

- Negative feedback reduces the effect of the initial change, for example:

Global warming increases evaporation and cloud cover

More solar energy is reflected, creating cooler temperatures

Typical mistake

Do not assume that a glacier will instantly respond to seasonal changes in the budget. There may be a time lag of many years between changes and glacial movement.

Glacials are periods of colder climate. These are separated by warmer periods called **interglacials**. Within the major glacial/interglacial cycles there are short-term fluctuations. Colder fluctuations are called **stadial** periods. Warmer periods are **interstadial**.

Typical mistake

Do not assume that the name negative feedback means it is 'bad' for the glacial system. It is called negative because it is reducing the impact of the original change.

14 Why can a glacier be classified as an open system?
15 What is the glacial budget?
16 What is the difference between positive and negative feedback?

Answers on p. 169

Climate change and the glacial budget over different timescales

REVISED

Causes of climate change through the Quaternary Ice Age

The Quaternary period is divided into two geological epochs:
- Pleistocene — 2.5 million to 11,500 years ago. This period spanned glacial and interglacial conditions, including the most recent glaciation
- Holocene — 11,500 years ago to today, which is the present interglacial period

The primary causes of climate change are the long-term changes in the Earth's orbit around the Sun. The **Milankovitch theory** links three orbit characteristics, which combine to minimise the amount of solar radiation reaching Earth:
- Eccentricity — the elliptical orbit changes to more circular and back again over a period of 100,000 years, varying the amount of solar radiation reaching Earth.
- Axis tilt — varying from 21.8° to 24.4° over a period of 41,000 years, thus changing the amount of solar radiation at the poles.
- Wobble — Earth wobbles on its axis over a 21,000-year cycle, changing the time of year when it is closest to the Sun.

Causes of change in the glacial budget through historical time

Changes may be due to:
- variations in energy from the Sun related to sunspot activity. Sunspots are temporary, cooler areas of the Sun's surface. An increase in sunspots can reduce the amount of solar energy reaching Earth, resulting in colder temperatures
- volcanic eruptions ejecting massive amounts of ash, sulfur dioxide, carbon dioxide and water vapour, which are spread around the globe by high-level winds, reducing the amount of solar radiation reaching Earth ('global dimming')

The **Little Ice Age** was a glacial oscillation between 1350 and 1850.
- Changes in the **thermohaline circulation** blocked the Gulf Stream to Europe, causing cooling of the climate.
- Increases in atmospheric pollution due to the Industrial Revolution may have resulted in climatic warming, ending the Little Ice Age.

Thermohaline circulation is the movement of ocean currents due to differences in temperature and salinity.

Exam tip

Remember that the start date of the Little Ice Age is an estimate — it can vary from the fourteenth to the sixteenth century. Climatologists agree that it depends on local conditions.

Seasonal changes and their impact on the glacial budget

Glaciers usually have a positive mass balance in winter and a negative one in summer, mainly due to seasonal changes in temperature. A series of colder than average winters may change the annual net budget.

> **Exam tip**
>
> You may be asked in the exam to analyse tables of figures representing the glacial budget. Make sure that you can calculate means of glacial mass balance measurements.

Glacier movement

REVISED

Differences between cold-based and warm-based glaciers

Table 1.20 **Characteristics of cold-based and warm-based glaciers**

Cold-based glaciers	Warm-based glaciers
• Occur in high latitudes — polar regions • Ice temperature below **pressure melting point** • Basal ice is frozen to bedrock • Most movement by internal deformation • Little erosion due to lack of movement	• Occur in temperate regions, for example Alps and Rockies • Temperature at base at pressure melting point • Heat from Earth adds to melting • Meltwater assists movement of glacier • Actively eroding and transporting

> **Exam tip**
>
> Some glaciers can be **polythermal** — cold-based in the upper region and warm-based lower down.

Glacier ice movement

- When **shear stress** overcomes friction, a glacier moves downslope.
- Velocities of glaciers vary greatly — 3 to 300 metres per year.
- Fastest flow is just below the surface in the centre of the glacier due to friction at the sides and base (Figure 1.14).

> **Pressure melting point** is a temperature below 0°C at which the ice can melt due to pressure from the weight of the ice.
>
> **Shear stress** is the downslope force due to gravity resulting from the build-up of an ice mass.

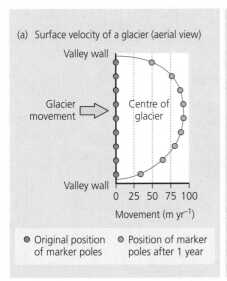

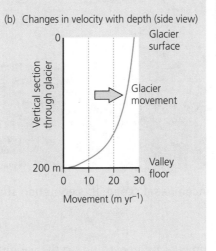

Figure 1.14 **Glacier velocities**

The movement from the zone of accumulation to the zone of ablation prevents ice build-up, keeping the glacier in a state of equilibrium.

Factors influencing the rate of glacier movement are explained in Table 1.21.

Table 1.21 Factors influencing the rate of glacier movement

Factor	Explanation
Gradient	Steeper slopes cause faster movement
Altitude	Affects temperature and precipitation inputs, which influence accumulation and the amount of meltwater
Lithology	Affects processes at the base and the level of friction
Size	Greater mass has greater potential velocity
Mass balance	Affects the equilibrium, influencing glacial advance or retreat

Glaciers can move by internal deformation, basal sliding and subglacial bed deformation.

Typical mistake

A retreating glacier will still have ice moving forward downslope. It just does not reach as far downslope as previously, before melting.

Internal deformation

This is how cold-based glaciers move, but it also occurs in warm-based glaciers. It comprises:

● **intergranular flow** — under pressure the ice crystals move relative to each other
● **laminar flow** — the ice crystals move along parallel layers within the glacier

Basal sliding

This applies to warm-based glaciers because it requires meltwater at the base to act as a lubricant between debris in the glacier base and the bedrock. It comprises:

● **enhanced basal creep** — basal ice deforms around irregularities in the bedrock surface
● **regulation slip** — basal ice deforms under pressure caused by obstacles; once past the obstacle the meltwater refreezes (Figure 1.15)

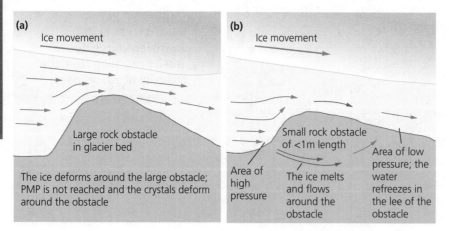

Figure 1.15 Basal sliding: (a) enhanced basal creep; (b) regelation slip

Subglacial bed deformation

Softer rock and unconsolidated sediments are not strong, so the weight of the ice in a glacier can cause the sediments to deform. As the sediments change shape, the ice on top moves with them.

Glacial surge conditions

Every 10–20 years, conditions can create a short-term period of movement up to 100 times faster than usual.

Meltwater builds up under the glacier, plus an increase in ice accumulation.

↓

Weight of ice prevents meltwater draining away and there is more due to the pressure melting point.

↓

Meltwater raises basal ice from bedrock, lubricating it, and allowing ice to flow more freely (surge).

↓

Surge releases meltwater and the glacier subsides on to bedrock, returning to normal flow.

Compressional/extensional flow

- **Compressional flow** — a reduction in gradient results in a slowing of movement. The ice thickens, crevasses close and thrust faults develop in the ice.
- **Extensional flow** — an increase in gradient results in accelerated movement. The ice thins and crevasses form (Figure 1.16).

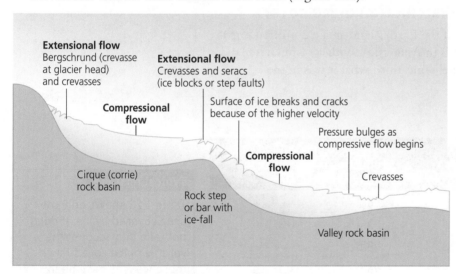

Extensional flow
Bergschrund (crevasse at glacier head) and crevasses

Extensional flow
Crevasses and seracs (ice blocks or step faults)

Compressional flow

Surface of ice breaks and cracks because of the higher velocity

Pressure bulges as compressive flow begins

Compressional flow

Crevasses

Cirque (corrie) rock basin

Rock step or bar with ice-fall

Valley rock basin

Figure 1.16 Compressional and extensional flow

Now test yourself

TESTED

17 What is the pressure melting point?
18 What are the differences between the types of movement for cold-based and warm-based glaciers?
19 What influences the rate of movement of glacial ice?

Answers on p. 170

The range of glacial environments and their distribution

Types of ice mass

Ice masses are classified according to their morphology, size and location. The main types of glacier can be subdivided into different types, as shown in Table 1.22.

Table 1.22 Different types of ice mass

Ice mass	Description	Size
Ice sheet	An ice dome, several kilometres thick, submerging the topography beneath	Larger
Ice shelf (sea ice)	A large area of floating glacier ice extending from the coast	↑
Ice cap	A smaller version of an ice sheet covering an upland area	
Ice field	Ice covering an upland area, but not burying topography	
Valley glacier	A glacier confined between valley sides	
Piedmont glacier	A valley glacier that fans out over a flatter area at the end of the valley	↓
Cirque glacier	A small glacier filling a hollow on the side of a mountain	Smaller

Past and present-day distribution of ice cover

- Today ice covers about 10% of the Earth's land area (15 million km^2).
- 85% of all current glacier ice is in Antarctica, with 10% in Greenland.
- Ice cover today is approximately one-third what it was in the Quaternary Ice Age (Table 1.23).

Table 1.23 Changes in ice cover

Region	Present area (thousands of km^2)	Estimated percentage reduction in size since the Quaternary Ice Age
Antarctica	1,350	6.9
Greenland	180	23.4
Arctic Basin	32	98.0
Asia	12	96.9
Alaska	5	99.7
Andes	3	96.6
European Alps	0.4	90.0
Scandinavia	0.4	99.9
Great Britain	0	100.0
Rest of the world	3.2	99.8

Typical mistake

Table 1.23 shows only the area covered by ice. It does not take into account the volume of ice cover. Much more ice may have been lost from Antarctica and Greenland than suggested by the figures for reduction.

Exam tip

Make clear in your answers if you are writing about an area that is still influenced by active glaciation or an area that was glaciated a long time in the past, leaving landforms as a result.

During the Quaternary Ice Age, ice sheets in North America and Europe were over 3 km thick. There was significant ice growth in southern South America, South Island New Zealand, Siberia and the Himalayas (p. 34).

Processes of glacial weathering, erosion and the characteristics and formation of associated landforms and landscapes

REVISED ☐

Glacial weathering

Freeze–thaw weathering involves repeated freezing and thawing of water, expanding cracks in rocks and eventually causing fragments to break off and fall on to the glacier.

Glacial erosional processes

Table 1.24 Glacial erosional processes

Process	Description
Abrasion	Debris embedded in the glacier base scrapes the bedrock as it moves.
Plucking (quarrying)	Ice freezing on to valley sides, floor and bedrock pulls away rocks as it moves.
Subglacial fluvial erosion	Meltwater flowing at the base of a glacier erodes rock the same way as surface streams. Pressure causes streams to flow faster, increasing erosion potential.

Factors affecting glacial erosion

- **Basal thermal regime** — warm-based glaciers are more actively moving, increasing their erosion potential.
- **Ice velocity** — faster-moving glaciers can be more erosive.
- **Ice thickness** — thicker ice increases pressure and erosion potential.
- **Bedrock permeability** — influences the amount of meltwater at the base, which affects the amount of plucking and subglacial fluvial erosion.
- **Bedrock jointing** — plucking occurs more readily in well-jointed rocks.
- **Debris characteristics** — large quantities of angular debris can increase abrasion.

Macro-scale glacial erosion landforms and landscapes

Cirques, arêtes and pyramidal peaks

A **cirque** forms when snow accumulates on a sheltered mountainside.

Freeze–thaw weathering and the removal of debris by meltwater creates a **nivation hollow**, which aids glacier formation.

Abrasion and plucking occur as the glacier moves in a rotational manner, deepening the hollow, steepening the backwall and creating a rock lip.

After glaciation the cirque may fill with water, forming a tarn.

Exam tip

Freeze-thaw weathering is also known as frost shattering.

Exam tip

Remember that freeze–thaw weathering will be much more active in areas where the temperature frequently rises above and falls below freezing, than it will in areas that are continually at sub-zero temperatures.

Typical mistake

Do not refer to freeze–thaw weathering as a type of erosion.

Typical mistake

Do not write generally about glacial erosion — be specific about the process.

A **cirque** is a bowl-shaped depression on a mountainside, also known as a corrie or cwm. Red Tarn on the side of Helvellyn in the Lake District is a lake formed in a cirque.

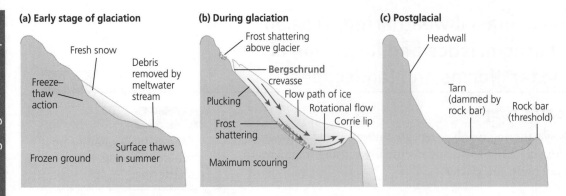

Figure 1.17 The formation of a cirque

- When two cirques erode headwards towards each other a steep, narrow ridge called an **arête** is formed, for example Striding Edge in the Lake District.
- A **pyramidal peak** is a steep-sided mountain peak formed when three or more cirques back on to each other. The Matterhorn in the Alps is a famous example.

A **bergschrund** is a deep crevasse near the backwall of a cirque. Meltwater flows into it, aiding glacial movement.

Glacial trough

As a glacier flows along a mountain valley, erosion processes deepen and widen it, changing it from V- to U-shaped. Gleann Einich and Glen Avon in the Cairngorms are examples.

Ribbon lake

Differences in bedrock and/or the rate of erosion create deeper hollows called **rock basins**. After glaciation the hollows flood to form long ribbon lakes, for example Loch Ness.

Revision activity

An exam question could use an OS map extract to locate glacial landforms in the UK. Make sure that you can recognise how a feature would appear on a map and can locate it using grid references.

Hanging valley

A smaller tributary glacier meeting a main valley glacier has less power to erode a deep valley, so it is above the level of the main valley. After glaciation a waterfall occurs. Yosemite Falls in Yosemite National Park, California is an example, falling 739 metres from a hanging valley.

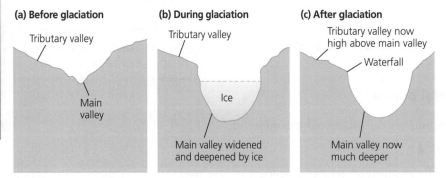

Figure 1.18 The formation of a hanging valley

Truncated spurs

A glacier cannot follow the turns around interlocking spurs in a river valley. It straightens the valley, forming cliff-like truncated spurs, for example the Devil's Point in Upper Glen Dee in the Cairngorms.

Meso-scale glacial erosion landforms and landscapes

Rôche moutonnée

An area of resistant rock on the valley floor is not completely removed. Under the ice, the up-valley side is eroded smooth by abrasion. The down-valley side is jagged due to plucking. Mt Iron near Wanaka in New Zealand is a 250 m-high rôche moutonnée.

Crag and tail

When a glacier is forced to flow around a large, resistant rock outcrop, the softer rock down-valley is protected. This results in a steep-sided up-valley slope (crag) and a gently sloping down-valley slope (tail). Edinburgh Castle is built on top of a crag-and-tail feature.

Micro-scale glacial erosion landforms and landscapes

Striations

Striations are scratches on the bedrock, parallel to the direction of ice movement, caused by abrasion.

Revision activity

Complete an internet search to find images of the named examples of landforms, so that you can picture the shape and size of each feature.

Now test yourself   TESTED

20 What affects the rate of glacial erosion?
21 What are the processes involved in the formation of a cirque?
22 What is the difference between a rôche moutonnée and a crag-and-tail feature?

Answers on p. 170

Processes of glacial and fluvioglacial transport and deposition, and the characteristics and formation of associated landforms and landscapes

REVISED

Glacial and fluvioglacial transport

A moving glacier **entrains** and carries material in three ways:
- **Supraglacial** — debris from weathering falls from valley sides on to the glacier.
- **Englacial** — debris falls into crevasses and is moved within the glacier.
- **Subglacial** — basal ice freezes around material and drags it along by **traction**. Englacial material moving to the base and plucking adds to the amount.

As well as small particles, a glacier can also transport large boulders long distances (e.g. **erratics**). Ice is the poorest sorter of transported sediments.

Exam tip

Remember meso (medium) and micro features are usually found within macro features such as glacial troughs.

Exam tip

Make sure that you can name an example for each of the different types of erosional feature.

Revision activity

Produce annotated diagrams, that could be used in your answers, to explain the formation of different glacial erosion landforms.

Entrainment is the process by which surface sediment is incorporated into the glacier.

Erratics are large boulders transported and deposited in areas of different geology. The Norber erratics in the Yorkshire Dales are good examples.

Glacial deposition landforms and landscapes

Glacial till

There are three types of **glacial till**:

- **Ablation till** — unsorted, angular material deposited by melting ice. Stones show no preferred orientation.
- **Lodgement till** — rounded, subglacial material deposited by a moving glacier. The long axis of stones is orientated in the direction of movement.
- **Deformation till** — weak bedrock is deformed (e.g. by folding) by ice movement.

> **Glacial till** is unsorted sediment carried by a glacier. A **moraine** is a deposit of glacial till.

Moraines

Table 1.25 Types of moraine

Moraine	Description
Terminal	A high ridge across a valley deposited as a glacier retreats from the furthest point reached.
Recessional	A series of ridges across a valley behind a terminal moraine, marking a stationary period in ice retreat.
Lateral	Weathered material falls on to a glacier from valley sides. When the ice melts it deposits a ridge parallel to the valley sides.
Medial	Two lateral moraines combine along the centre of the glacier surface when valley glaciers merge. As ice melts it is deposited along the middle of the valley.
Push	When glaciers begin advancing again, the debris at the snout is pushed into a ridge.

Drumlins

Lodgement till is moulded into small, egg-shaped mounds with the long axis orientated in the direction of flow. The up-valley slope is steeper than the slope facing down-valley. Drumlins found in concentrations or 'swarms' are referred to as **basket-of-eggs topography**.

Now test yourself

TESTED ☐

23 What is the difference between a terminal moraine and a push moraine?

Answer on p. 170

Fluvioglacial processes

- Meltwater can be supraglacial, englacial, subglacial and proglacial (beyond the ice front).
- Discharge is highest in late spring and summer, each day reaching a mid-afternoon peak.

Meltwater erodes, transports and deposits sediment (Table 1.26). Material deposited by subglacial meltwater is called an **ice-contact deposit**. Material deposited down-valley from the glacier is a **proglacial** or **outwash deposit**.

Table 1.26 Fluvioglacial processes

Fluvioglacial process	Features
Erosion	• Proglacial meltwater erosion is similar to that in normal streams • Subglacial streams are under pressure and fast flowing, eroding bedrock especially by abrasion
Transportation	• High-energy meltwater streams have the capacity to transport large sediment loads

Deposition	• When meltwater loses energy, it deposits material that is:
	  – rounder than glacial deposits
	  – sorted by size, with the heaviest deposited first
	  – stratified into distinct layers

Fluvioglacial landforms — ice-contact features

Esker

● Long, sinuous ridge of sorted material on the valley floor (Figure 1.19).
● Deposited when the flow of subglacial meltwater in tunnels decreases.
● The Thelon Esker in Canada is 800 km long.

Kame

● Conical, flat-topped hill of stratified material.
● Meltwater washes material into crevasses or depressions in the glacier surface, which is deposited when the ice melts.

Kame terrace

● Flat-topped ridge along the side of a glacial valley.
● Meltwater streams flow along the sides of a glacier. When the flow is reduced, material is deposited.
● Strath Nethy in the Cairngorms is an example.

Exam tip

Another theory of esker formation is that they represent a rapidly retreating delta that formed as ice melted.

(a) During glaciation

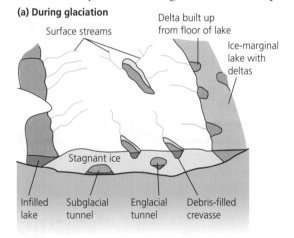

Surface streams
Delta built up from floor of lake
Ice-marginal lake with deltas
Stagnant ice
Infilled lake
Subglacial tunnel
Englacial tunnel
Debris-filled crevasse

(b) After glaciation

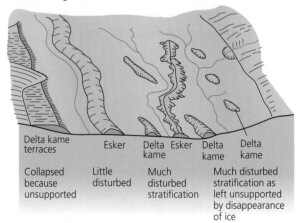

Delta kame terraces
Esker
Delta kame
Esker
Delta kame
Delta kame

Collapsed because unsupported
Little disturbed
Much disturbed stratification
Much disturbed stratification as left unsupported by disappearance of ice

Figure 1.19 The formation of eskers and kames

Proglacial features

Sandur (outwash plain)

- Gently sloping, sorted layers of sand and gravels.
- Meltwater streams lose energy, entering a lowland area and depositing their load. Coarser materials are deposited first and finer materials furthest from the glacier.
- Sandurs are found in many areas, such as the glaciated areas of Iceland.

Varve

- An annual deposit of material when meltwater enters a glacial lake.
- In spring and summer meltwater increases, transporting and depositing heavier material.
- In winter there is less meltwater, so finer sediment settles out.

Kettle hole and lake

- Small, shallow depression or lake.
- Blocks of ice become detached and covered in sediment in the outwash plain.
- When the ice melts the deposits collapse, leaving a hollow, which can flood to form a lake.

> **Exam tip**
>
> Kame deposits are often found with kettle holes in areas known as **kame and kettle topography**, for example Kincaid Park in Anchorage, Alaska.

Now test yourself

TESTED

24 What is the difference between ice-contact and proglacial deposits?

Answer on p. 170

Suites of landforms within glacial landscapes

REVISED

Variations between highland and lowland glacial landscapes

Table 1.27 Characteristics of highland and lowland glacial landscapes

Highland glacial landscapes	Lowland glacial landscapes
- Some areas of glaciers - Characterised by ice erosion features, including source features such as cirques (corries), arêtes and pyramidal peaks - Landscape eroded by valley glaciers, producing features such as glacial troughs and truncated spurs - Depositional features such as moraines can be found - Distinct patterns can be seen, for example 71% of corries in the northern Highlands of Scotland face north and east, where accumulation is greater on shaded slopes - Areas such as the Cairngorms, Lake District and Snowdonia show the characteristics of highland glacial scenery	- Some areas of slow-moving or stagnant ice - Major areas of glacial and fluvioglacial deposition features - Orientation of drumlins and deposited rock fragments align with the direction of flow of glaciers and ice sheets - Include proglacial features - The landscape near Telford in East Shropshire shows evidence of depositional features

Variations between ice sheet and valley glacier landscapes

Table 1.28 Characteristics of ice sheet and valley glacier landscapes

Ice sheet landscapes	Valley glacier landscapes
• Glacial erosion over a large area known as **areal scour** • Low-lying, ice-smoothed hills containing meso-scale erosion features • Cold-based ice sheets produce little erosion. Areas of scattered erratics can be a feature • Depositional features where the ice sheet is stagnant or slow moving • As the edges of ice sheets re-advance and retreat, the features can be complex • Cover large areas, such as southern Canada and northern USA	• Erosion processes are concentrated in valleys, forming glacial troughs • Meso- and micro-scale erosion features are found within the valley • Glacial and fluvioglacial depositional features are found, especially moraines • Periglacial features are present where the valley sides are not ice-covered • Valley will be modified by fluvial erosion and weathering after the glacier retreats • The Lake District (e.g. Langdale) shows evidence of the action of valley glaciers

Typical mistake

Do not assume that ice sheets are a highland feature. They can be found in lowland areas.

Areas of areal scour can also be referred to as **knock and lochan** topography, after areas in northwest Scotland.

Exam tip

When writing about the characteristics of different glacial landscapes, do not just list the landforms, but show awareness of the scale and overall character of the landscape.

Revision activity

For each of the four types of glacial landscape described in Tables 1.27 and 1.28, name an example of the landscape and outline examples of the glacial features found in it. Use the internet to research what each type of landscape looks like.

Periglacial processes and the formation of associated features

REVISED

Periglacial processes result from:
- daily temperatures below 0°C for at least 9 months and below −10°C for 6 months per year
- frequent cycles of freezing and thawing as temperatures fluctuate during the year
- precipitation below 600 mm per year

Periglacial refers to the edges of glacial areas, where repeated freezing and thawing modify the landscape.

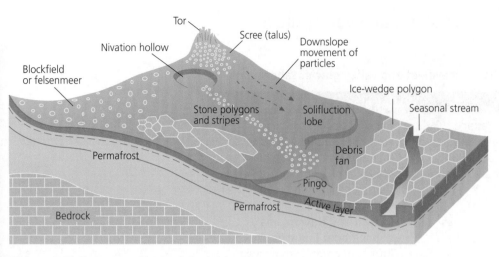

Figure 1.20 Features of a periglacial landscape

Ground ice formation and associated features

The cold temperature freezes the ground to form **permafrost**, which can be:

- **continuous** — a layer of frozen ground that can be hundreds of metres deep
- **discontinuous** — a thinner, fragmented layer of frozen ground
- **sporadic** — an isolated mass of permafrost in unfrozen ground

The **active layer** is the surface layer up to 3 m deep, which thaws in summer and refreezes during the winter.

Ice lenses

- Moisture accumulates and freezes in soil and rock pores.
- As ice grows it wedges the rock apart, causing **frost heaving**.

Ice wedge polygons

- Soil contracts and cracks in extreme cold.
- Water fills the cracks in spring and refreezes to form **ice wedges**.
- These build up over the years. Viewed from the air, they form polygonal patterns.

Patterned ground

- Frost heave expands soil upwards, causing surface stones to run down the slope created.
- Over time the stones form circles and polygons. On steeper slopes the circles become elongated, forming stripes.

> **Permafrost** is soil and rock that is below 0°C for at least 2 years.

> **Typical mistake**
>
> Permafrost does not form quickly after 2 years of cold temperatures. Most of the existing permafrost was formed during previous glacial periods and has remained during warmer interglacials.

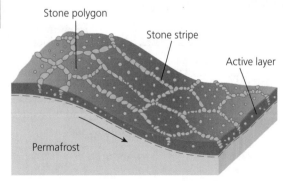

Figure 1.21 Features of patterned ground

Pingos

Two possible methods of **pingo** formation are outlined in Table 1.29.

Table 1.29 Methods of pingo formation

Closed system	Open system
Unfrozen lake over permafrost prevents thawing ↓	Areas of discontinuous permafrost in valley bottoms ↓
When the lake drains or fills with sediment the insulation is lost and the waterlogged lake bed freezes ↓	Water soaks into upper layers of ground and flows to low-lying areas ↓
Hydrostatic pressure results in newly frozen water accumulating into ice lenses, pushing up the lakebed sediments	Water freezes, creating an ice core which grows as more water flows in, forcing up the upper layers of sediment

Thermokarst landscape

Thermokarst features are the result of melting permafrost.

Frost heave-formed domes and ice lenses collapse, leaving depressions that fill with water.

Frost weathering and mass movement features

Nivation hollows

When snow accumulates in a sheltered area, freeze–thaw weathering weakens the rock beneath it. Meltwater erodes and transports the debris away, leaving a depression.

Blockfield and scree slopes

In flat areas freeze–thaw weathering and frost heave produce areas of angular boulders. On slopes and cliff faces, **mass movement** causes the weathered rock fragments to fall. These build up on the lower slope or cliff base to form a **scree slope**.

Pro-talus ramparts

Debris falling from a cliff face on to a steep snow slope slides to the foot of the slope, accumulating into a mound.

Solifluction terraces

The melting active layer creates a saturated layer of soil, which moves downslope to form a terrace. Small contoured steps can form along the slope to create terracettes, while on steep slopes a 'tongue-like' lobe forms.

Asymmetric valleys

Differing rates of solifluction due to valley orientation can result in one side of the valley being steeper than the other.

Head deposits

Fragmented, weathered material is deposited due to solifluction, often in valley bottoms.

Pingos are dome-shaped mounds of earth up to 70 m high and 500 m in diameter, with an ice core.

Relict landforms were formed under past climatic conditions and by processes very different from those found today.

Mass movement is the downward movement of materials due to gravity.

> **Exam tip**
>
> Remember that thermokarst may be a **relict** landscape made up of features formed by past climatic conditions and processes very different from those found today. However, it can also be produced currently by human activity melting permafrost (p. 51).

> **Exam tip**
>
> Remember that freeze–thaw weathering is not unique to periglacial areas, but it can have a much greater impact there.

Periglacial action of water and wind and associated landforms

- Most erosion by water occurs in the warmer seasons when the active layer thaws.
- Many valleys are **dry valleys**, cut by meltwater in the past.
- Many meltwater streams are braided due to sediment load.
- Lack of vegetation due to limited water availability for plant growth allows wind erosion.
- Fine sediments are blown from outwash plains and deposited as **loess**, forming large, flat areas of rich soils — **loess plateaux**.
- Material along both sides of parts of the Mississippi River comprises periglacial loess deposits.

Now test yourself
TESTED

25 What is meant by the term 'relict feature'?
26 What does the term 'periglacial' mean'

Answers on p. 170

Variations in glacial processes, landforms and landscapes over different timescales
REVISED

Glacial processes vary over a timescale ranging from seconds to millennia (Table 1.30).

Table 1.30 The timescale of glacial processes

Timescale		Example of change
Fast	Seconds	Rapid mass movement
		Rapid glacier melt
	Seasonal	Changes in accumulation and ablation
		Meltwater flows
		Changes in depth of the active layer
	Decades/centuries	Changes in the glacial budget
	Millennia	Stadials and interstadials
		Glacials and interglacials
Slow		Post-glacial working of deposits

Changes in seconds

- **Rapid mass movement** — rockfalls and landslides change the profile of a glacial valley.
- **Rapid glacier melt** — volcanic activity causes large-scale melting, resulting in flooding and rapid mudflows (**lahars**).

Seasonal changes in meltwater discharge

- Increase in discharge allows the transportation of more and larger material on, in and under the ice.
- Streams emerging from glaciers have more erosion potential.
- Decreasing discharge only allows transportation of small sediment.
- Changes result in **varve** deposits.

Changes over millennia

Weathering, mass movement and fluvial processes rework glacial landscapes:

- Glacial troughs contain scree and become asymmetrical.
- Glacial lakes and kettle holes become infilled with sediments.
- Runoff reworks moraine deposits.
- Reduced discharge produces a **misfit** stream, which cuts a meandering channel in the glacial valley floor.

Glacial processes are a vital context for human activity

Impact of glacial processes and landforms on human activity

- Climate change resulting in less ice leads to a decrease in available meltwater, which may be an important water supply for a country's needs.
- A sudden release of meltwater that has accumulated within a glacier or on the surface behind ice or a moraine dam can cause devastation. These **glacial lake outburst floods** (jökulhlaup) are caused by:
  - a rise in water level caused by an increase in ice flotation
  - melting of an ice dam due to geothermal heat (p. 35)
  - failure of a moraine dam

> **Exam tip**
>
> Glacial lake outburst floods (GLOFs) can release millions of cubic metres of stored water at flow rates of thousands of cubic metres per second, resulting in impacts over vast areas away from the glacier where the meltwater had collected. As well as being an example of how human activity can be affected, they provide a contrasting scale of feature.

Impacts of human activity on glacial processes and landforms

Humans can have a direct impact through the exploitation of resources and indirectly through human-created climate change.

Extraction of sands and gravels

Outwash deposits sorted into sand and gravels are an important source of raw materials for the construction industry.

Creation of reservoirs

Dams are built in glacial valleys to create reservoirs so that hydro-electric power (HEP) can be produced year-round.

Tourism

Tourists can be attracted to glacial areas for many reasons, such as the dramatic scenery and to take part in activities such as walking (including walking on glaciers), climbing and skiing.

The impact of tourism can result in development in remote areas and damage to fragile landscapes and ecosystems. For example, the Athabasca glacier on the Columbia icefield in the Canadian Rocky Mountains is easily accessible by road. 100,000 vehicles a month drive past the glacier in summer and over 800,000 visitors a year visit the glacier, many using special 'Snowcoach' tour buses to drive on to the glacier surface. The area has become crowded and developed, with car parks and a large visitor centre with facilities for tourists. Problems include litter attracting wildlife such as bears into the areas, while walking off paths is damaging the fragile ecosystem of the area.

Management strategies

There are six strategies that can be adopted (Table 1.31).

Table 1.31 Glacial area management strategies

Management strategy	Description
Do nothing	Area can be exploited by many economic activities
Business as usual	Leaves management as it is, so may include some environmental policies
Sustainable exploitation	Some economic development with mandatory environmental regulation
Sustainable development	Develops the area, allowing the use of resources for the present population while maintaining resources for future generations
Comprehensive conservatism	Protects and conserves the environment — the only economic activity might be ecotourism
Total protection	No access to the environment allowed, except for research

Exam tip

Make sure that you have detailed, up-to-date knowledge of *one* located management strategy involving *either* the impact of glacial processes and landscape on human activity, *or* the impact of human activity.

Revision activity

If using the Athabasca glacier and Banff National Park as examples, search for images of these places on the internet to help you understand the impact human activity has had on these areas.

Case study: Strategy to manage the impacts of human activity on glacial processes and landforms and landscapes

Banff National Park, Canada

- A UNESCO World Heritage site and Canada's first national park.
- Covers 6,641 km² of mountainous terrain and alpine landscapes, including glaciers and ice fields.
- The park receives 5 million visitors a year and millions more pass through on the way to other destinations.
- There are 3,600 hotel rooms within the park as well as other accommodation, such as campsites. There is pressure for more development to cater for tourism.
- Visitor pressure has had an impact on wildlife, including loss of habitat. In the past wildlife was seen as a threat and pest.
- Management:
  - The town of Banff has been limited to a population of 10,000 people and development strictly controlled.
  - An airstrip and cadet camp near the town, which prevented animal movement, have been removed.
  - Plans to increase the size of golf courses have been refused.
  - Controls have been placed on the size of winter skiing areas.
  - Wildlife crossings for wolves and bears have been constructed over the Trans-Canadian Highway to reduce accidents and to reduce the impact of habitat fragmentation.
  - Bison have been reintroduced into the ecosystem.
  - Prescribed fires are used to mimic natural fires, which help maintain biodiversity.
  - Increased visitor awareness that they are in a national park and World Heritage site has been achieved by signage and information points. These also raise awareness of the importance of the wilderness area, its wildlife and its habitats.
  - Interpretative information boards linking glacial processes, landforms and historic tree lines are used to raise awareness of climate change.

Permafrost degradation through human activity

The melting of permafrost can result in a **thermokarst** landscape characterised by depressions and hummocks as the ground settles unevenly. It can be caused by anything that depletes the layer of surface vegetation, for example deforestation, agriculture and construction.

Removal of snow cover or vegetation removes insulation, allowing increased thawing, while buildings and infrastructure also raise the temperature of the ground.

Permafrost thaw results in ground subsidence, which can damage buildings, roads, railways and bridges. Solutions include:

- buildings built on piles to raise them 1 m above the permafrost
- roads built on thick layers of gravel, which acts as insulation, protecting the permafrost
- utilities such as water and energy delivered in insulated pipes above the ground

The Batagaika crater in East Siberia is an example of a large thermokarst depression — 1 km long and 100 m deep — formed as a result of the surrounding deforestation.

Exam practice

The format for the different examination papers is shown below.

Specification	Method of examination
Eduqas A-level Component 1; Section A	Two compulsory data-response questions and one extended-response question
Eduqas AS Component 1; Section A	Two compulsory data-response questions; an extended response is required in part of each question
WJEC AS Unit 1; Section A	Two compulsory data-response questions

Eduqas A-level format

1 Study Figure 1, which shows the coastline at Barton-on-Sea in Hampshire.

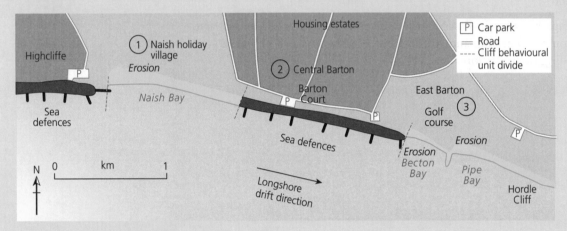

Figure 1 The coast at Barton-on-Sea

a) Use Figure 2 to describe the possible factors that may have influenced the decision to build sea defences in this area. [6]

b) Suggest why the sea defences in Figure 1 may impact on the coast further east at Hurst Spit. [5]

c) Outline what is meant by the term 'managed retreat'. [2]

2 Table 1 shows historic rates of erosion of cliffs at two points along the coast of eastern England.

Table 1 Coastal erosion rates at two locations in eastern England

	Location A				Location B			
	Top of cliff		Foot of cliff		Top of cliff		Foot of cliff	
Date	Total (m)	Average (m/yr)	Total (m)	Average (m/yr)	Total (m)	Average (m/yr)	Total (m)	Average (m/yr)
1975–1980	10.4	2.08	7.60	1.52	12.4	2.48	13.3	2.67
2010–2015	1.0	0.2	0	0	16.5	3.30	17.3	3.46

a) Use Table 1 to compare rates of erosion at Location A and Location B. [6]

b) Outline one way in which sub-aerial processes influence the shape of a cliff. [2]

c) Suggest how future changes in sea level might influence rates of coastal erosion at the locations shown in Table 1. [5]

WJEC AS format

3 a) Use Table 1 to compare rates of erosion at Location A and Location B.   [5]
   b) Outline one way in which sub-aerial processes influence the shape of a cliff.   [3]
   c) Examine the success of one strategy to manage the impact of human activity on coastal
       landscapes.   [8]

Eduqas A-level format

4 Assess the success of one strategy to manage the impact of human activity on
   coastal landscapes.   [15]

Eduqas A-level and WJEC AS format

5 Study Table 2, which shows the results of a survey of the shapes of glacial and fluvioglacial
   sediments.

Table 2 The shapes of glacial and fluvioglacial sediments

Shape of sediment	Sediment location		
	Supraglacial debris (on the glacier)	Subglacial debris (under the glacier)	Fluvioglacial debris (in meltwater)
Very angular	31%	–	–
Angular	69%	25%	2%
Sub-angular	–	42%	6%
Sub-rounded	–	33%	68%
Rounded	–	–	24%
Very rounded	–	–	–

a) Use Table 2 to analyse the variations shown in sediment shape.   [Eduqas A-level 6; WJEC AS 5]
b) Outline the physical processes that result in the accumulation
    of subglacial debris.   [Eduqas A-level 2; WJEC AS 3]
c) Explain how the process of fluvioglacial deposition leads
    to the formation of distinctive landscapes.   [Eduqas A-level 5; WJEC AS 8]

Eduqas A-level format

6 Figure 2 shows the present distribution of permafrost.

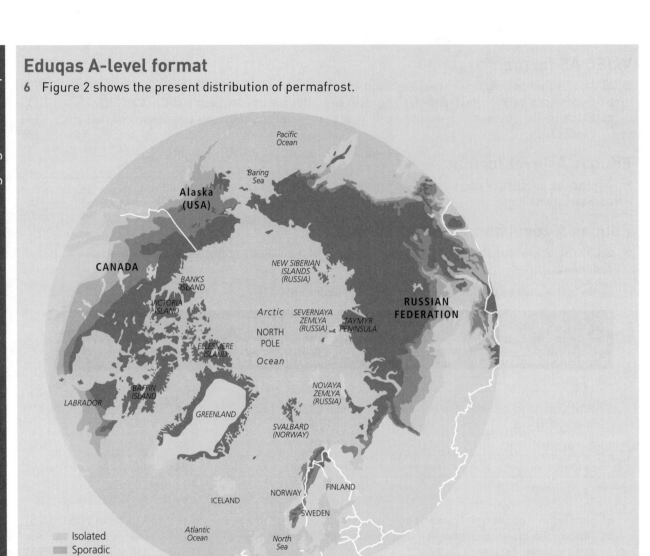

Key:
Isolated
Sporadic
Discontinuous
Continuous

Figure 2 Present distribution of permafrost

a) Use Figure 2 to compare the distribution of continuous and discontinuous permafrost. [5]
b) Explain the formation of two features associated with periglacial areas. [6]
c) Outline one way in which human activity can degrade permafrost. [2]
7 Assess the success of a strategy to manage the impact of human activity on glacial landscapes. [15]

Answers and quick quiz 1 online

ONLINE

Summary

You should now have an understanding of the following:

Coastal landscapes

- The coastal zone is an open, dynamic system.
- The coastline can be divided into cells based on the movement of sediment.
- Coasts are influenced by processes operating on very different timescales.
- High-energy coasts produce environmental landforms, while low-energy coasts produce depositional landforms.
- Physical factors such as waves, tides and geology can influence the development of coastal features.
- The combination of marine erosion processes and sub-aerial processes results in the formation of distinctive coastal landforms.
- The processes of transportation and deposition result in distinctive low-energy coastal landforms.
- Aeolian, fluvial and biotic processes can form their own distinctive landforms in coastal environments.
- Coastal processes can have both positive and negative impacts on human activity.
- Human activity can have both positive and negative impacts on the coastal environment.

Glacial landscapes

- A glacier is an open system.
- The glacial budget changes over both long-term and short-term periods.
- Glaciers move in a variety of ways. Basal temperatures and other factors influence the type and rate of movement.
- Ice masses occur on a range of scales. Coverage has decreased since the Quaternary Ice Age.
- The combination of glacial erosion and weathering results in the formation of landforms on a range of scales.
- The processes of glacial and fluvioglacial transportation and deposition result in the formation of distinct landforms and landscapes.
- Periglacial processes produce distinctive landforms at the edges of glacial areas.
- Glacial processes and landforms can have an impact on human activity.
- Human activity, including management, can impact on glacial landforms and periglacial processes.

2 Changing places

Changing places

Relationships and connections

The characteristics of places

Many factors affect the character of a **place**, such as:

- **demographic** — population size and structure
- **socio-economic** — employment and education opportunities, income, health, crime rates
- **cultural** — religion, customs, social behaviour and language
- **physical geography** — relief, features such as rivers, coasts and geology
- **location** — urban or rural, relationships with other settlements
- **built environment** — land use, building type and density
- **political** — role of government and council, for example planning regulations

The character of places can be shaped by the following:

- Flows of people:
  - Migrants moving in to live or work can result in areas of a town being influenced by the culture of the migrant population, such as shops and services catering for their needs. These influences may be at a very local scale.
  - People travelling to another area for work. On a small scale, villages surrounding a large town may become commuter settlements with limited services for the population as they rely on the nearby large settlement.
- Flows of resources:
  - For example, reliance on raw materials from another location.
  - The import of cotton resulted in the growth of the cotton industry in Lancashire, with urban areas characterised by mills. The decline in imported cotton and the decline in the cotton industry has led to changes in the character of the region.
- Flows of money:
  - Foreign direct investment (FDI) from a company based elsewhere can influence the character of a region by creating employment and wealth, such as Nissan's £100 million investment in Sunderland.
  - Ending investment in a place can change a town or region's character, giving an impression of decline, which can discourage future investment. The end of investment by Tata Steel into the Redcar Steelworks on Teesside is an example of the economic character of a town changing quickly.
  - Investment by **MNCs**, especially fast-food chains, has led to a decrease in the uniqueness of some places. Town centres frequently become homogenised as 'clone towns', with over 60% of shops being chain stores. The impact can also be global, with identical styles and products being recognisable in many countries.

> **Exam tip**
>
> You *must* support your studies in this section with examples from contrasting places. You should use your 'home' place, which may be a locality, neighbourhood or small community, as well as other regional or national examples. You could also use examples from field studies.

> **Connections** refer to any types of physical, social or online linkages between places.
>
> **Place** is a portion of geographic space to which meaning has been given by people, and is best understood as a small settlement or a neighbourhood within a larger city.
>
> An **MNC**, or multinational corporation (sometimes referred to as a multinational company), is a business that has factories and/or offices in more than one country. An MNC often has a centralised head office in its home country from where it coordinates management of its global facilities.

- Flows of ideas:
  - Urban planners and developers change a place, with redevelopments encouraging people to move back into previously run-down and derelict parts of a city, such as many dockland and waterfront areas in UK cities.
  - Ideas coming from universities can lead to area becoming a technology hub, such as Cambridge.

How continuity and change can affect lives

The factors that shape the character of a place can continue to influence it for a long time. For example, West Ham United Football Club was located in Upton Park, east London because it evolved from the Thames Ironworks club at the start of the twentieth century. This gave continuity to the area long after the industries had disappeared. However, the club's recent move to the Olympic Stadium (now called the London Stadium) in Stratford, east London shows that factors can change, as it consequently changed the character of Upton Park, which is being redeveloped for the building of 850 new homes.

Continuation or change in any of the factors could impact on the following aspects of our own or other people's lives:
- employment opportunities
- access to services and open space.
- factors affecting health (e.g. atmospheric pollution)
- social inclusion or isolation
- ability to afford housing

Changing places — meaning and representation

REVISED

Perceptions of a place

Different people have varying perceptions of a place based on information from two sources:
- **direct experience** — based on living in a place
- **indirect experience** — based on information seen or read in the media or other sources

How a place is given meaning results from a person's perceptions of, engagement with and attachment to the place. These can be influenced by a range of factors (Table 2.1).

> **Typical mistake**
>
> Do not think of 'place' as meaning an urban area. It could be part of a city or a remote rural location, including wilderness areas.

> **Revision activity**
>
> In what ways has your own life been affected by the factors and flows that connect where you live with other places?

> **Exam tip**
>
> The factors shown in Table 2.1 will influence your perception of a place. Keep an open mind and use any evidence presented before you make a judgement.

Table 2.1 Factors influencing the perception of a place

Factor	Example
Age	An older person's perception may be different from that of a young person — a person's needs change with age, altering perceptions (e.g. suitability based on access to services)
Gender	Males may perceive a place or individual street as more or less friendly/safe than females do
Socio-economic status	A wealthy person may have different perceptions of a 'cheaper' housing area than the residents living there
Socio-cultural positioning	Ethnic origin, race and religion influence how a person perceives the role of a place or its suitability in terms of services and amenities

Residents of a place perceive it differently from visitors, especially somewhere like a tourist resort. Even visitors may have different perceptions of the same place, for example due to the weather at the time of the visit.

A person may have an emotional attachment based on memories and feelings about events connected with the location. These feelings may be positive or negative.

Exam tip

Remember that a local person may have a perception of a place that is based on factors that no longer influence its character. A visitor or newcomer may consider such factors no longer important.

Revision activity

For your home area or a place you know, compile a list showing:
a) what factors give it a distinct 'character'
b) varying ways in which it is perceived by different groups of people

Variety of ways in which places are represented

A place can be represented in both **informal** and **formal** ways (Table 2.2), which can influence perception:
- Formal place representations are produced by political, social and cultural agencies (including local government, education institutions, tourist boards and national heritage agencies), along with large businesses.
- Informal place representations are produced by individuals or small groups of people working outside of formal-sector institutions. Informal representations are often creative and do not necessarily try to faithfully reproduce reality.

Table 2.2 **Examples of ways a place can be represented**

Formal		Informal		
		Factual	Non-factual	Opinion
News media	Census data	Media written by individuals or informal groups	TV dramas	Social media
Photographs	Statistics		Films	TV
Advertising	Geospatial data		Literature	Graffiti
Promotional materials (e.g. by councils/tourist organisations)	Maps			

Place meanings can have an effect on continuity and change for places, and can affect the lives of people

The ways in which different groups perceive and give meaning to a place lead to variations in the demand for changes to a place, which could in turn lead to conflicts. For example, Table 2.3 shows the demands for change that might occur in a rural area such as a national park.

Revision activity

Compile examples showing how places have been informally represented in different ways.

Table 2.3 Variations in place meanings and their impact on continuity and change

Group	Perception of the place	Desire for continuity or change
Local residents	Attractive place to live	Development of infrastructure for modern life, such as high-speed internet Improved public transport Limited new housing development
Young people	A place with limited opportunities	Increased employment opportunities Affordable housing and reduction in second home ownership High-speed internet and good mobile coverage
Farmers	A place of livelihood	Development of modern agricultural practices Maintaining the environment
Visitors	An area of attractive scenery	Preservation of the scenic landscape and villages

Changes in attachment to an area have positive or negative impacts on people. For example, a terrorism incident changes the perception of safety of a place, quickly reducing tourist numbers. This impacts on employment opportunities and the income of residents.

Now test yourself

TESTED

1 How do places differ in terms of their social, economic and environmental characteristics?
2 What are the factors that can influence a person's perception of a place?

Answers on p. 170

Changes over time in the economic characteristics of places

REVISED

Economic change can lead to structural changes in employment

Over time, economies go through changes that affect the **employment structure**. The modified Clark-Fisher model identifies three stages (Figure 2.1).

> **Employment structure** refers to how the workforce is divided between the four main employment sectors: primary — the collection of natural resources; secondary — manufacturing; tertiary — services; and quaternary — research and development activities.

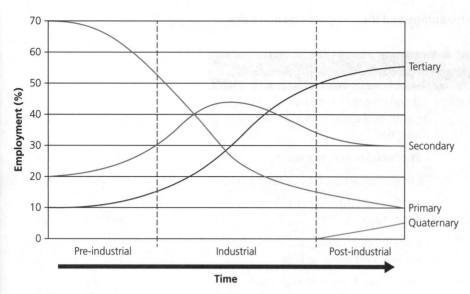

Figure 2.1 The modified Clark–Fisher model

Table 2.4 The characteristics of changing economic stages

Economic stage	Description
Pre-industrial	Primitive economies with the primary sector dominating, for example **LICs** such as Mali
Industrial	Growth in secondary and decline in primary activities; some increase in tertiary activities, for example emerging economies such as China
Post-industrial	Primary activities remain low and secondary activities decline, while tertiary activities dominate; there is a growth in quaternary activities, for example **HICs** such as the UK

The World Bank defines **LICs** and **HICs** as follows:
- LIC (low-income country) — GNI per capita below $1,005
- HIC (high-income country) — GNI per capita over $12,235

The term **emerging economy** has been used by geographers and economists to refer to middle-income countries.

Globalisation is the way in which economies and societies have become integrated by a global network of trade, communication and immigration.

- Such economic developments change the employment opportunities in the different economic sectors, which increases the need for retraining.
- The impact of **globalisation** may alter the sequence or speed of development.
- The tourism industry may result in missing out a large industrial stage.

Forces and factors influencing economic restructuring

A place is influenced by factors that result in changes to the employment structure:
- **Changes in technology** — machinery reducing the need for farm labourers; robots reducing the need for skilled secondary workers.
- **Depletion of resources** — natural resources are depleted, resulting in a decline in the extractive industry.
- **Changing lifestyles and tastes** — a decline in demand for products such as coal, and a reluctance to work in old, heavy industries.

Typical mistake

Do not assume that all countries follow the Clark–Fisher model of development. Remember also that within one country different regions could have very different employment structures.

- **Government strategies** — policies encourage the growth of new industries in declining areas.
- **Globalisation** — large MNCs, with access to cheaper raw materials and labour, and with economies of scale, create competition, which leads to industrial decline in a region.

The decline in primary employment in rural areas and in secondary employment in urban places — deindustrialisation

The employment structure of the UK has changed greatly since the nineteenth century (Figure 2.2).

> **Deindustrialisation** is the process of economic and social change in an area caused by a reduction in industrial activity or employment (sometimes due to automation).

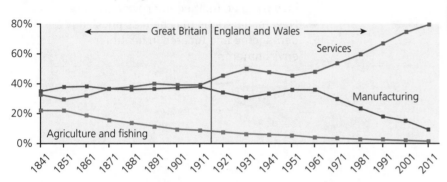

Figure 2.2 Changes in the UK employment structure 1841–2011

Since the late 1970s the UK has seen:
- a significant decline in employment in the secondary sector
- a smaller decline in primary activities
- an increase in, and domination of, the tertiary sector
- a growth in the quaternary sector

Table 2.5 Changes in UK employment 1978–2017

Economic activity	Employment in June 1978	Employment in December 2017
Agriculture, forestry and fishing	520,000 (1.9%)	453,000 (1.2%)
Mining and quarrying	382,000 (1.4%)	64,000 (0.2%)
Manufacturing	6,732,000 (24.9%)	2,689,000 (7.6%)
Tertiary	16,098,000 (59.7%)	26,184,000 (74.6%)
Professional, scientific and technical (quaternary)	965,000 (3.5%)	3,056,000 (8.7%)
Other (e.g. construction)	2,291,000 (8.6%)	2,655,000 (7.7%)

Source: Office for National Statistics

> **Typical mistake**
>
> Do not assume that deindustrialisation means less is produced. In the UK, manufacturing production has continued to grow at an average of 0.7% a year despite declines in secondary sector employment.

Rural areas have experienced employment changes too: these include primary sector job losses linked with coal mine closures and farming modernisation. There are positive and negative impacts on the lives of people in affected rural areas (Table 2.6).

> **Exam tip**
>
> Remember that the impacts of deindustrialisation can be classified as economic, social and environmental.

Table 2.6 Impacts of deindustrialisation on people in rural mining areas

Negative impacts	Positive impacts
<ul><li>Decrease in employment opportunities</li><li>Changes in rural communities due to rural–urban migration in search of jobs</li><li>Ageing population remains</li><li>Positive feedback results in declining services</li><li>Environmental concerns over industrial dereliction in rural mining areas (e.g. South Wales)</li><li>Increasing isolation (especially in farming)</li><li>New jobs on short-term or zero-hours contracts</li></ul>	<ul><li>Reduction in environmental pollution</li><li>Potential new employment opportunities due to industrial heritage tourism</li><li>Removal of old industrial buildings, creating a better-quality environment in which to live</li><li>Potential new leisure and recreation facilities (e.g. flooded gravel pits)</li><li>Land available for new housing, easing the housing shortage</li><li>Import of cheaper goods allows disposable income to go further, improving welfare</li><li>New tertiary jobs may be less physically demanding and located in healthier environments</li></ul>

Now test yourself

TESTED

3 What is globalisation?
4 What is deindustrialisation?

Answers on p. 170

Economic change and social inequalities in deindustrialised urban places

REVISED

Consequences of the loss of traditional and secondary industries in urban areas

- Many of the impacts of deindustrialisation in rural areas also apply to urban areas.
- The major impact is unemployment due to a lack of new job opportunities.
- These impacts are greater when an urban area relies on one secondary industry employer, and may be felt for many years.
- For example, in the 1980s, in Consett, northeast England, 4,300 people out of a population of 30,000 were directly employed by the steelworks, with three times that number indirectly employed by it. Closure of the works pushed the male unemployment rate to almost 100% for a time.

The cycle of deprivation

Unemployment resulting from deindustrialisation can create a 'spiral of decline' in urban areas (Figure 2.3). This is another example of positive feedback (p. 9).

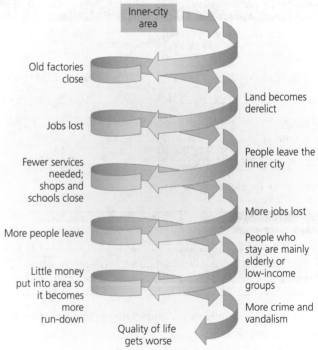

Figure 2.3 The inner-city spiral of decline

Deindustrialisation and the spiral of decline can result in a cycle of deprivation, which impacts on people in urban areas (Figure 2.4).

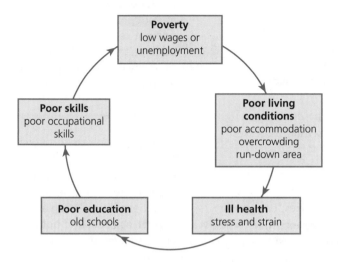

Figure 2.4 The cycle of deprivation

Table 2.7 Potential impacts of deindustrialisation on quality of life

Quality of life indicator	Impact of deindustrialisation
Income	Lack of income due to unemployment
	New jobs may be lower paid, part-time or based on zero-hour contracts
Employment	Limited job opportunities
	New jobs require retraining or are unskilled
Health	Declines due to stress-related illness, and poor diet and living conditions
	Reduction in atmospheric pollution improves aspects of health
Education	Access to good schools may be limited; in areas with long-term unemployment, education may not be considered worthwhile
	Limited opportunities to learn new skills for employment
Access to housing and services	Lack of income limits access to housing, much of which is unaffordable
	Positive feedback (here, a vicious circle) results in services closing as people cannot afford them
Crime rate	Studies show that areas with high levels of deprivation are significantly more likely to suffer higher crime rates, including violence, especially among young people, as a way of acquiring status among peers
Living environment (indoor and outdoor)	Poverty means only the poorest-quality housing is available
	The poorest housing areas are frequently in areas with high traffic densities and poor air quality

Social exclusion

Social exclusion can result from unemployment due to deindustrialisation and the cycle of deprivation:

- Excluded groups are unable to take part in activities due to cost.
- They cluster together in socio-economic areas in an urban area based on affordability — a type of **spatial segregation**.
- Non-excluded individuals move away from the area, isolating the excluded group further.
- Areas where excluded groups cluster go into decline due to lack of money for maintaining or improving the area, which increases exclusion.

Pollution levels

- Industrial decline, especially in old traditional industries, can result in a decline in atmospheric pollution (Table 2.8). For example, when the steelworks closed in Consett, County Durham in 1980, there was a noticeable decrease in air pollution. As a result, gardeners found that pests started to damage plants for the first time.

Table 2.8 Changes in atmospheric pollution levels 1970–2016

Pollutant	Decrease 1970–2016
Sulfur dioxide	97%
Nitrogen oxides	72%
Non-methane volatile organic compounds	66%
Particulate matter	Over 70%

Source: Defra

- Deindustrialisation of old inner-city areas can result in decreases in traffic congestion, reducing pollution levels.
- Modern tertiary industries require less energy usage and produce less atmospheric pollution.
- A reduction in noise pollution accompanied the decrease in old manufacturing processes.
- Old industrial buildings can be removed or improved, and waste tips landscaped, reducing visual pollution of the environment.

Government policies in deindustrialised places

Enterprise zones

An **enterprise zone** is an area where policies are used to encourage economic growth and development.

Enterprise zones were first used in the 1980s and again after 2012 as part of long-term economic plans, with favoured sites being those with little or no existing business occupancy. Benefits include:

- business rate relief and financial benefits
- simplified planning rules
- infrastructure for business — superfast broadband, transport links and a location with co-businesses

There are 48 enterprise zones in England, including Birmingham and Manchester. By 2015, English enterprise zones had created 19,000 jobs and attracted 540 companies and £2.2 billion of private investment.

Social exclusion is the inability to participate in the normal relationships and activities available to the majority of people. It can be economic, social, cultural or political.

Exam tip

Remember that there are many factors that result in social exclusion, such as gender or race, and not just economic decline. It may result from a combination of factors. Social exclusion occurs in rural areas often without the development of spatial segregation.

Typical mistake

Do not make the assumption that the decrease in pollution is all due to deindustrialisation. Changes in fuel and regulations controlling emissions are other contributors.

Revision activity

Research present-day and historic images of deindustrialised places to see how the areas have changed. Suggest how these changes will impact on the lives of people living there.

Exam tip

Policies are continually changing and evolving, especially in the present-day economic climate. Keep up to date with events and policies that influence the areas you have studied.

Exam practice answers and quick quizzes at **www.hoddereducation.co.uk/myrevisionnotesdownloads**

Local Enterprise Partnership (LEP)

- LEPs were established in England in 2011 as voluntary partnerships between local authorities and businesses to promote economic growth and job creation in local areas.
- There are 38 LEPs in England, such as the CIOS (Cornwall and the Isles of Scilly) LEP, whose partnership with the European Space Agency is investing £8.4 million of Local Growth Fund into Goonhilly Earth Station. This will create the world's first commercial deep-space communications station, capable of tracking future missions to Mars and the Moon.
- Since 2011, LEPs have supported 196,000 businesses, creating 180,600 jobs. They have established £7.6 billion of private investment, helped build 93,200 homes and supported 217,900 learners.

Retraining

Assistance is available to help with retraining and learning a new career. Funding can be in the form of:

- grants and bursaries from an organisation
- professional and career development loans — the government pays the interest while a person studies
- advanced learner loans covering the costs of a training course

In 2018 the National Retraining Partnership between the government, CBI and TUC was set up to establish a National Retraining Scheme to tackle skills shortages in new economic growth activities.

EU Growth Programme

The **European Structural and Investment Funds** 2014–2020 provide investment for innovation, businesses, skills and employment. There are three types of fund:

- **European Social Fund** — aims to improve employment opportunities, promote social inclusion and invest in skills.
- **European Regional Development Fund** — supports research and innovation, small-to-medium-sized enterprises and development of a low-carbon economy.
- **European Agricultural Fund for Rural Development** — helps rural businesses to grow and expand.

The 'Brexit' referendum of 2016 means that these funds may cease to be available in the UK in the future.

Foreign investment by multinational corporations (MNCs)

FDI mostly takes the form of investments by large MNCs, often based in the EU, USA, Japan, China and India (e.g. India's Tata has acquired the Port Talbot steelworks).

- In 2016/17 there were 2,265 projects in the UK, creating over 75,000 new jobs.
- Enterprise zones help encourage FDI in the UK.
- MNCs aim to gain some benefit from their investment, such as access to a market, avoiding trade barriers and lower labour costs.
- Not all FDI is in deindustrialised areas.
- Many new jobs are in the tertiary or quaternary sector (e.g. workers for US MNC Amazon).

> FDI (foreign direct investment) is cross-border investment made by residents and businesses from one country to another, with the aim of establishing a lasting interest in the country receiving investment.

Answers on p. 170

Now test yourself

TESTED

5 What is the cycle of deprivation?
6 How is the closure of shops sometimes an example of positive feedback?

The service economy (tertiary sector) and its social and economic impacts

REVISED

The tertiary sector developed to support secondary industry and the needs of workers. Today, 79% of the UK's GDP comes from, and 80% of workers are found in, the service sector — an example of **tertiarisation**.

Expansion of retailing, commerce and entertainment in central areas

These tertiary activities originally clustered in central areas for the following reasons:

● There was a large catchment area population to support the services.
● These were the most accessible points in the past.
● Access to a large workforce.
● New technological infrastructure would develop in these areas first.

Expansion of the tertiary sector has occurred because:

● many people have higher incomes to spend on basics and more disposable incomes to spend on services and luxuries
● the ageing population may be mortgage-free and have more disposable income
● modern technology has created a demand for new services
● changing tastes have led to the growth of certain retail services, such as coffee shops

> **Tertiarisation** is when the service sector comprises the biggest element of the economy.
>
> **Re-urbanisation** is the movement of people and economic activities back into city centres.
>
> **Gentrification** is the process by which a place changes from being a poor area to becoming a richer one.

Gentrification and associated social changes in central urban places experiencing re-urbanisation

People have been moving back into cities for the following reasons:

● Old industrial areas have been redeveloped, turning derelict industrial buildings into apartments. For example, Little Kelham in Sheffield is on a former steelworks site, and involved the building of new homes as well as the conversion of old industrial buildings into apartments.
● People unable to afford property elsewhere, including rural areas, move to cheaper areas near city centres.
● Increasing pressure due to rapid population growth in urban areas forces people into central areas.
● Increasing numbers of young professionals prefer to live in the central area to be close to services such as restaurants and entertainment. The number of 20- to 29-year-olds living in city centres such as Manchester has tripled since 2001.

Re-urbanisation can result in **gentrification**:

● High-income, young professionals can afford to renovate older housing.

Exam practice answers and quick quizzes at **www.hoddereducation.co.uk/myrevisionnotesdownloads**

- Old industrial buildings are converted into luxury accommodation rather than affordable social housing.
- Services develop in the area to cater for the higher-income population.
- As the area becomes more desirable, it attracts other wealthier people into it.
- Tredegar Square in Mile End, London was built in the 1830s, comprising grand town houses. By the 1970s the area was run down, with homes in a poor state of repair. Many had been made into small flats, let to poorer members of the population. By the 1990s, having been saved from demolition, with ownership transferred from big landlords to individuals, the housing stock had been renovated and the area became a desirable place to live for young, wealthy people.

Re-urbanisation can result in social change in an area:
- Low-income residents are priced out of buying or renting in the area. People in the bottom 10% of relative deprivation are the most affected.
- Members of the original population feel socially excluded.
- The ethnic diversity found in deprived areas changes as the new, incoming population dominates.
- An increase in higher-paid professional, managerial, technical and creative jobs attracts young professionals, excluding older and unskilled people in the area.

The impacts of changes in the service economy

Changes in the service economy are complex. While some areas are growing, others are in decline.

Features of cities facing decline include:
- higher rates of poverty
- lower rates of employment growth
- lower rates of immigration of economically active groups

Ten of the top 12 most declining cities in the UK are in northern England, for example Rochdale, Burnley and Bolton, reflecting the impact of deindustrialisation as well as geographic location.

The tertiary sector is not just found in central urban areas — it also operates in other locations (Table 2.9).

Table 2.9 Alternative locations of tertiary activities

Location	Description
Out-of-town retailing	Superstore or a collection of chain stores close to good transport links, for example Meadowhall in Sheffield
Office park (business park)	Office buildings grouped together, usually in the outskirts, which offer lower building costs, good transport links, less congestion and a pleasant environment, for example Cobalt Business Park in North Tyneside, which is the largest in the UK, employing 14,000 people
Leisure and entertainment complex	Leisure facilities such as cinemas are increasingly found in outer areas, often with retail parks Stadia are moving to redeveloped areas or the outskirts for more space, for example in 1998 Reading FC moved to the new Madejski Stadium, which was built on a landfill site outside the town
Home (internet shopping)	The increase in shopping online is having an impact on the tertiary services found in central areas and elsewhere

As well as bringing social changes, developments in the service sector can affect people in the following ways:

- Closure of shops and services in the central area, because they are unable to compete with other locations and the internet, can leave people feeling isolated.
- Closure of chain stores such as Mothercare, Toys R Us and House of Fraser has resulted in many people becoming unemployed. This can lead to even more closures in the area in a 'domino effect'.
- The central area becomes characterised by poorer-quality shops and less choice (e.g. the growth of charity shops).
- Movement to out-of-town locations favours those with cars.

> **Exam tip**
>
> Do not just write about over-general 'change' in places. Refer to specific processes, such as re-urbanisation, gentrification or changes in economic activity.

Now test yourself

TESTED

7 What is the link between re-urbanisation and gentrification?

Answer on p. 170

The twenty-first century knowledge economy (quaternary sector) and its social and economic impacts

REVISED

The **knowledge economy** is associated with:

- high-tech manufacturing (computers, electronics and aerospace)
- science-sector industries such as education, healthcare, software design and biotechnology
- business services such as insurance, information and communications

Despite modern communications such as broadband, quaternary industries in the knowledge economy tend to cluster together. Companies that have located on science parks have had higher growth rates.

> The **knowledge economy** is an economy based on creating, evaluating and trading ideas and information. Essentially, this is a synonym for the quaternary sector.

Factors encouraging cluster growth

Table 2.10 Factors encouraging clustering in the knowledge economy

Factor	Description
Proximity to universities and research establishments	• The government allows universities to establish local growth plans and University Enterprise Zones (e.g. Bristol and Nottingham) with business spaces for new high-tech companies in the early stages of development • Companies require a highly educated workforce; often there are links to a university, providing research and ideas, for example Cambridge • Strong research in UK universities attracts foreign investment, especially from MNCs
Government support	• In 2017 the government announced a 50% increase in research and development by 2027 • Government bodies such as the Medical Research Council support research • Establishment of enterprise zones encourages new industries • East London Tech City (Silicon Roundabout) in Shoreditch, London has received local and national government support; MNCs such as Google have also invested, and London universities are academic partners
Planning regulations	• Local Enterprise Partnerships can grant automatic planning permission for certain developments, such as new industrial buildings, within specified areas
Infrastructure	• Industry locates in areas with good connectivity to other parts of the country and abroad; they require high-speed broadband links as well as proximity to good road, rail and air links

The impacts of quaternary industry clusters on people and places

- Areas such as rural locations with limited communication, especially broadband, will not benefit from the growth of quaternary industries (Figure 2.5).
- Clusters attract a higher-educated, digitally proficient workforce earning higher salaries, which changes the social character of a place.
- Clusters lead to improvements in **global connectivity**, which attracts further interest and investment in a place as well as increasing productivity in research. This can lead to the development of a **global hub** (p. 102).
- Job opportunities for less skilled workers are limited.
- Higher-paid workers demanding homes results in house price rises.
- Services concentrate on catering for quaternary workers.
- Non-quaternary workers suffer social exclusion.
- Modern science/business parks in pleasant surroundings with good communications help in 'place-making', creating a good image and attracting new workers and investors.

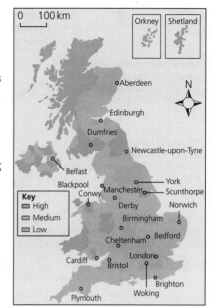

Figure 2.5 Potential areas of digital exclusion in the UK

Recently, clusters have developed in redeveloped or gentrified inner-city areas, which are popular with young professionals. East London Tech City (Silicon Roundabout) in Shoreditch, London grew because rates were reasonable for technology start-ups and the recession of 2008 made many professionals redundant from investment banks, creating an available, skilled workforce, A number of notable technology companies, such as Google, Facebook and Amazon, are active in the cluster.

The growth of the cluster could push up rents in the area, which could limit start-ups and the growth of new companies.

> ### Now test yourself
> TESTED
>
> 8 What are the advantages of clustering for a knowledge economy business?
> 9 How can a quaternary industry cluster have a negative effect on people?
>
> **Answers on pp. 170–1**

The rebranding process and players in rural places

REVISED

Rebranding, **regeneration** and **re-imaging** of urban and rural places have become prominent in areas experiencing economic decline and deindustrialisation.

The **post-productive countryside** faces a number of economic and social challenges (Figure 2.6).

> **Exam tip**
>
> When using a case study, make it clear that it is an example of rebranding, regeneration or re-imaging, or a combination of more than one.

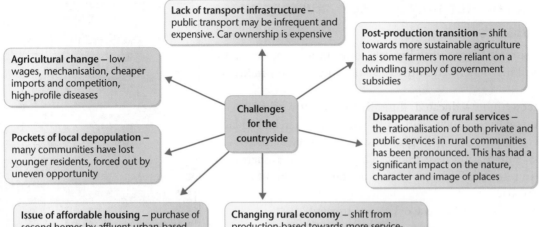

Lack of transport infrastructure – public transport may be infrequent and expensive. Car ownership is expensive

Agricultural change – low wages, mechanisation, cheaper imports and competition, high-profile diseases

Post-production transition – shift towards more sustainable agriculture has some farmers more reliant on a dwindling supply of government subsidies

Challenges for the countryside

Disappearance of rural services – the rationalisation of both private and public services in rural communities has been pronounced. This has had a significant impact on the nature, character and image of places

Pockets of local depopulation – many communities have lost younger residents, forced out by uneven opportunity

Issue of affordable housing – purchase of second homes by affluent urban-based middle-class householders. This form of gentrification inflates prices, excluding low-income would-be buyers

Changing rural economy – shift from production-based towards more service-oriented employment. There has been fragmentation of the local economy, creating uncertainties. Fewer guaranteed jobs

Figure 2.6 Challenges facing rural areas

The perception of rural areas can be influenced by what is seen in the media. Some areas are also digitally excluded, lacking both mobile phone coverage and fast internet connections.

Rural rebranding (Figure 2.7) could be based on a number of themes (Table 2.11).

Table 2.11 Potential rebranding themes for rural areas

Rebranding theme	Examples
Recreation	Increasing access to areas by establishing walking/cycling routes, such as the mountain bike trail opened in Swaledale in the Yorkshire Dales in 2018, or festivals, for example the Isle of Wight Walking Festival
Heritage	Encouraging an interest in historic buildings and the past/industrial past, for example Morewellham historic village, river port and copper mine in Devon
Media	Using a location used in a TV series or film to attract visitors, for example Highclere Castle in Hampshire, the setting for the TV series *Downton Abbey*
Event management	Establishing events based on crop/food festivals, for example the Great British Food Festival at Hardwick Hall, Derbyshire or using land for music festivals to attract visitors, such as the Latitude Festival in Suffolk
Food and produce	Advertising specialist or artisan foods from the area, e.g. Yorkshire has developed a number of trails based on a food (e.g. cheese) or drink (e.g. beer or gin) theme

Rebranding is the way a place is changed and marketed so that the image and perception of it is improved.

Regeneration is a long-term process aimed at improving the economy and social environment.

Re-imaging involves changing the reputation and perceptions of a place through specific improvements.

Post-productive countryside places more emphasis on sustainability, with less intensive food production, in an attempt to limit the environmental impact of agriculture.

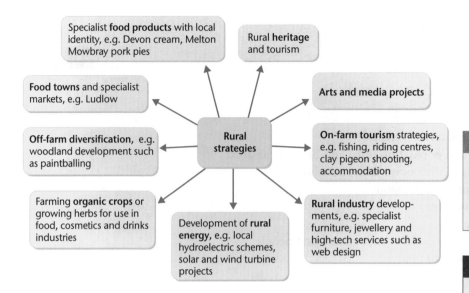

Figure 2.7 Strategies for rebranding rural areas

Rebranding may be initiated by formal and informal sector players working on different geographical scales:

- individuals or small groups of volunteers
- local businesses
- parish or local councils
- county councils
- National Park authorities
- tourist organisations
- media interest
- governments (UK, Welsh Assembly, Scottish Parliament) and the EU

Remember that rebranding may occur sometime after the activity has become established.

The effects of rebranding rural areas on the perceptions, actions and behaviour of people

- Perceptions of an area become more positive, increasing visitor numbers.
- Businesses catering for visitors benefit from increased trade.
- People move into the area, helping to maintain services.
- Development of a community 'identity'.
- New business and employment opportunities catering for increased visitor numbers/population, which diversifies the local economy.
- Investment improves the quality of the environment, making the area attractive for visitors.
- Increase in visitors results in traffic congestion and an environmental impact.
- Demand to move to the area raises house prices, excluding the local population.
- Increased second home ownership changes the character of villages.
- Local people may resent the influx of visitors and newcomers.

Revision activity

Research examples of rural rebranding to investigate how attempts have been made to create a 'brand'.

Typical mistake

Do not just learn the rebranding process that has occurred in a rural area. Make sure you know what the area was like *before* rebranding took place.

Exam tip

Make sure that you have named/located examples of different types of rebranding activity that have been initiated by different groups.

Typical mistake

Do not assume that rural rebranding is only to increase tourism. While many schemes are tourism based, some attempt to attract new businesses. The rebranding strategy will depend on what advantages the area is perceived to have.

Rural management and the challenge of continuity and change

Managing rural change and inequality in diverse communities

Changes in rural population characteristics include:

- a smaller proportion of the local population born and bred in the area
- an increase in people moving to the countryside, possibly to work from home or to retire (counterurbanisation)
- an increase in second home owners
- changing migrant worker numbers
- an ageing population due to increased life expectancy and people retiring to the country (Figure 2.8)

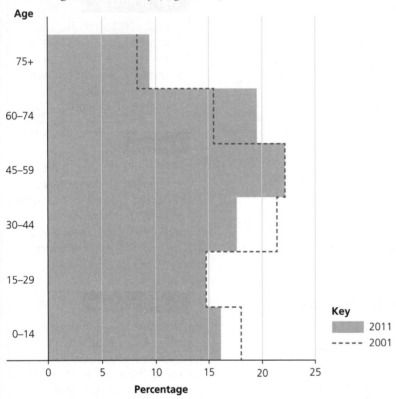

Figure 2.8 Changes in the age of the resident population of rural areas 2001–2011

- changes to the socio-economic characteristics of the population, with a proportion of locals in low-income jobs and newcomers in higher-income employment
- an increase in the number of people commuting to nearby urban areas for work

Housing issues

- In popular areas, counterurbanisation and demand for second homes raises house prices, making them unaffordable for locals.
- Second home ownership leaves villages deserted during the week or out of season.
- There is a reluctance to allow new housing developments in rural locations (**NIMBYism**).

Transport and service provision issues

- Rural areas have seen the removal of bus services because they are unprofitable.
- Services such as shops, pubs, banks and post offices close down due to costs.
- Mobile phone coverage can be poor and broadband speeds slow due to a lack of infrastructure, resulting in digital exclusion, especially as services move online. This may also deter affluent self-employed people from relocating to the countryside if the area lacks reliable broadband.
- Health service facilities are a long distance away, in the nearest urban area.
- Supermarkets and Amazon may not deliver to isolated areas, such as Scottish islands.
- Those who do not drive feel at a disadvantage and are isolated (mobility deprivation).
- There is increasing reliance on community-run services and increasing use of volunteers to maintain services.

Ongoing challenges in rural areas where regeneration/rebranding is absent

- Lack of investment in the area.
- Decline in employment opportunities.
- Lack of service provision.
- Out-migration, especially of the younger population.
- Decline in newcomers to the area.
- Need to retrain the original population to use modern technology.

> **Typical mistake**
>
> Do not assume that all rural areas that have not been rebranded are in decline, with a low-paid, unskilled, elderly population. Many rural areas have an active community with many highly skilled individuals.

Managing change in rural communities associated with counterurbanisation and second home ownership

- Limit the purchase of new-build homes to locals.
- Provide financial assistance to first-time buyers in rural areas.
- Use Community Land Trusts (CLTs) to manage and provide affordable housing.
- Allow planning controls that limit the use of second homes and allow development.
- Increase taxation on second homes (reduced council tax has already been removed).

The rebranding process and players in urban places

REVISED

Re-imaging and regenerating urban places

Urban areas can change their image in a variety of ways (Figure 2.9 and Table 2.12).

> **Exam tip**
>
> When using examples of rebranded areas, be specific and give the name of the area rather than just the name of the town or city. However, remember to say in which urban area it is found.

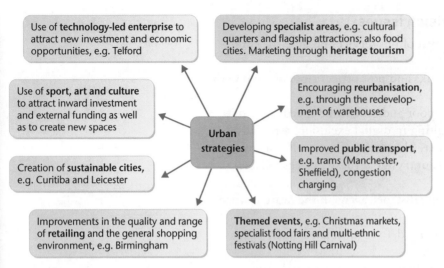

Figure 2.9 Strategies for rebranding urban areas

Table 2.12 Methods of re-imaging urban areas

Method of re-imaging	Description
Promoting advantages for business	Advertising global links, suitable buildings and potential lifestyle, for example Newport in South Wales promotes its good transport links, economically active workforce, lower staff costs and rents than London, and its investment in digital connections making it a 'super-connected city'
Sport / music stadia	Improvements linked to a major sporting or cultural event, for example hosting the 2012 Olympics became a catalyst for regeneration of run-down parts of the Lea Valley in east London
Cultural quarters	Promoting culture, such as museums and theatres, to increase visitor numbers Parts of an urban area might provide a distinctive cultural background, which becomes an attraction, for example the Balti Triangle in Birmingham
Festivals	Events such as food festivals, Christmas markets or cultural events (e.g. the Edinburgh Festival) boost tourism numbers each year
Industrial heritage	Restoration of industrial areas attracts visitors (e.g. the Leeds Industrial Museum at Armley Mills, 2 miles from the city centre) and/or the development of small craft industries
Flagship developments	A development encourages new businesses and tourist attractions, for example the Scottish Event Campus on the River Clyde in Glasgow

Revision activity

Research online to see how Newport has attempted to attract new businesses.

Investigate flagship developments to see what types of development are encouraged.

A **flagship development** is a high-profile land and property development providing a catalyst for further redevelopment.

Re-imaging and regenerating urban places through external agencies

Different agencies can be involved in urban rebranding, including:
- **government and local government** — provide the impetus and finance for large flagship developments

- **corporate bodies** — responsible for the implementation of a regeneration plan
- **community groups** — may be responsible for organising festivals
- **local councils** — may give backing to the rebranding theme and to festivals

The impact of urban re-imaging and regeneration

- Increase in employment opportunities for individuals.
- Individuals and/or businesses move to the regenerated parts of the urban area.
- Areas become popular visitor destinations for tourists or inhabitants from other parts of the urban area.
- The improved environment improves the quality of life for individuals.
- The nature of the area changes to cater for visitors, limiting services for the resident population.

> **Typical mistake**
>
> Do not assume that everybody will be in favour of rebranding. Changes in the neighbourhood, such as increased congestion and changes in services, may not be popular with some locals.

Urban management and the challenges of continuity and change

REVISED

Re-imaging and regenerating affect the social and economic characteristics of urban places

- Changing the perceptions of an area may lead to conflicting outcomes.
- Changing perceptions make a place considered safer to live, attracting people.
- Higher-earning, younger people move into regenerated areas, pricing out the local population.
- New homes as part of the regeneration may not be affordable for many locals.
- The original population may not have the skills for newly established jobs.
- The area becomes popular with visitors, creating problems such as noise and limited parking for residents.
- The perception of a newly regenerated area attracts investment, leading to decline in other parts of the urban area.

Ongoing challenges in urban places where regeneration and rebranding are absent, have failed or cause overheating

- In some urban areas, regeneration or rebranding has not occurred, while others are in danger of **overheating**.
- Areas with no regeneration continue to decline, or decline accelerates, as businesses move to newly regenerated areas.
- The decline results in kick-starting a cycle of decline and deprivation.
- Rebranding can fail because many inner urban areas are crowded and congested, and not suitable for modern industry, so little investment occurs.
- The urban area may be remote from growth areas, markets or good communications, resulting in the failure of regeneration.

> **Overheating** refers to when an economy has had a prolonged period of growth but the supply of goods cannot meet demand. In many cities, this is seen in the housing market.

These issues are difficult to solve.

Overheating can create its own challenges:

- The high demand for housing has become a serious problem, raising prices and pricing many people out of the market, especially young, first-time buyers. Even building large numbers of houses may have little impact on prices because demand is so high.
- Conflict over where to build new homes to meet the demand, such as greenfield or brownfield sites. 300,000 new homes a year are needed, which increases pressure to build on greenfield sites.
- Increasing pressure for urban areas to expand into green-belt areas.
- The need to increase service provision to meet growing population needs.
- Developing infrastructure, especially transport, to cope with the growth.
- Changes in the socio-economic characteristics. Parts of an urban area may become segregated by wealth, lifestyle, age or skill.
- Increase in immigration, creating more demand for homes and services as people move into the area at the prospect of good jobs and wealth.
- Changes in services to cater for the section of the population with a high level of disposable income at the expense of the less wealthy members of the population.

Now test yourself

TESTED

10 What are the negative impacts of rebranding a rural area?
11 What problems can be created by urban regeneration?

Answers on p. 171

Exam practice

The format for the different examination papers is shown below.

Specification	Method of examination
Eduqas A-level Component 1; Section B	Two compulsory data-response questions and one extended-response question
Eduqas AS Component 2; Section A	Two compulsory data-response questions; an extended response is required in part of each question
WJEC AS Unit 2; Section A	Two compulsory data-response questions

Eduqas A-level format

1 Figure 1 shows employment and deprivation in Greater Manchester.

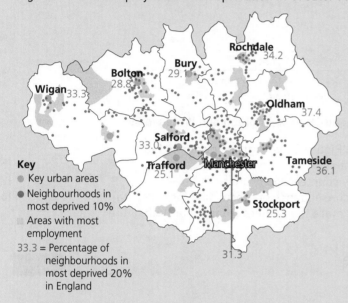

Key
- Key urban areas
- Neighbourhoods in most deprived 10%
- Areas with most employment
- 33.3 = Percentage of neighbourhoods in most deprived 20% in England

Figure 1 Employment and deprivation in Greater Manchester

a) Using Figure 1, describe the pattern of deprivation shown on the map.   [5]

b) Explain how government policies have assisted deindustrialised urban places such as Greater Manchester.   [8]

WJEC AS format

2 a) Using Figure 1, describe the pattern of deprivation in Greater Manchester.   [5]

   b) Explain one possible reason for the pattern of deprivation shown in Figure 1.   [3]

   c) Assess the success of one or more government policies that provide assistance to deindustrialised urban places.   [8]

Eduqas A-level format

3 Figure 2 shows the occupations of rural residents aged 16–74 in England and Wales in 2011.

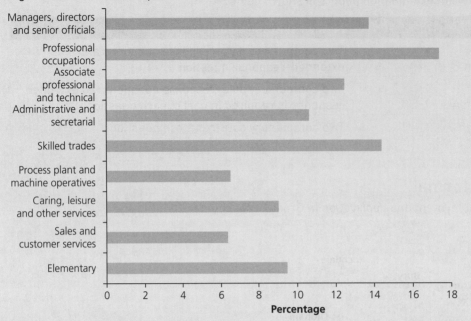

Figure 2 Occupations of rural residents aged 16–74 in England and Wales in 2011

a) Suggest how the occupations shown in Figure 2 may provide evidence for
   counterurbanisation in England and Wales.   [5]
b) Outline one social challenge created by counterurbanisation.   [2]
c) Explain one positive and one negative consequence of rebranding for rural places. [6]
4 Examine why some places have benefited more than others from the growth of the
   twenty-first century knowledge economy.   [15]

Answers and quick quiz 2 online

ONLINE

Summary

You should now have an understanding of the following:

- Many factors influence the character of a place.
- Different people have a different perception of a place due to a variety of factors.
- Places can be portrayed in a variety of ways, both formal and informal.
- Places change economically over time, which impacts on people in a variety of ways.
- Deindustrialisation has consequences for people and has resulted in government policies to encourage economic growth.
- The growth of the tertiary sector, re-urbanisation and gentrification have all had an impact on some urban areas.
- Quaternary industries frequently cluster, impacting upon urban areas.
- Rebranding rural areas creates challenges and affects the lives of people.
- Urban areas can be rebranded, with a variety of consequences.
- Place changes are often contested by different groups of people, meaning that tension and conflict often result, especially where there is overheating.

3 Global systems

Water and carbon cycles

The concepts of systems and mass balance

The water cycle as a system

- The global hydrological cycle (Figure 3.1) is a closed system.
- **Inputs** and **outputs** move to and from stores within the system. None crosses the system's boundaries.
- The amount of water in the system is fixed (1.38 million km^3), which means the **mass balance** does not change.
- Water is held in **stores** in, on and above the Earth.
- **Flows** are **transfers** of water between stores.

> **Exam tip**
>
> The water and carbon cycles are studied as systems to show the interrelationships between the land, oceans and atmosphere. You should have an understanding of systems from your study of coastal or glacial landscapes.

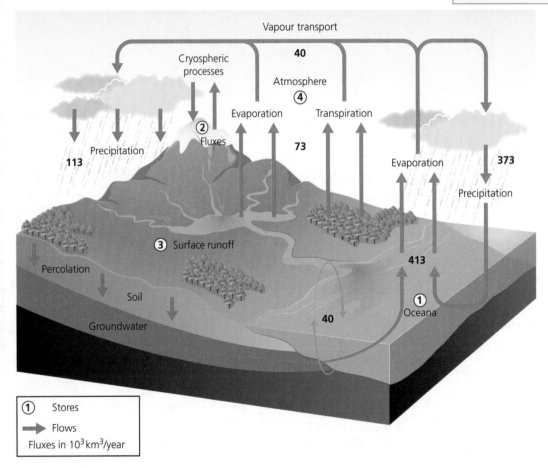

Figure 3.1 **The global water cycle**

The major water stores

Table 3.1 The world's major water stores

Store	Percentage of the world's water	Location
Saltwater stores	**97.0**	
Oceanic	97.0	Oceans and seas covering 71% of the Earth
Freshwater stores	**3.0**	
Cryosphere	1.9	Contains 68.7% of all fresh water
		Covers mainly high-latitude and high-altitude glacial areas, and includes **permafrost**
Groundwater and soil moisture	1.1	Below the surface
		Influenced by climate and geology
Terrestrial surface water	0.01	Rivers and lakes
		Distribution influenced by climate
Atmosphere	<0.01	Most in the form of water vapour around the globe
Biosphere flora and fauna	<0.01	Distribution influenced by climate

Changes in the size of stores

Sea-level change

- **Eustatic change** has resulted in sea levels changing significantly over geological time.
- During glacial periods more water is frozen, decreasing water in the oceans.
- Climate warming increases ice sheet melting, raising sea levels.

Changes in cryosphere processes

- Short-term changes in ice **accumulation** and **ablation** occur annually due to seasonal changes in temperature.
- Climatic changes, resulting in glacial and interglacial periods, cause significant changes in the size of the cryosphere (p. 94).
- Human-induced global warming may be permanently reducing the cryosphere by increasing ablation.

Processes that control transfers in the water cycle across a range of timescales

- Short-term storm events increase transfers locally.
- Seasonal variations in climate impact on transfer rates (e.g. monsoon climate).
- Climatic variability due to events lasting years such as **El Niño Southern Oscillation** impacts on precipitation levels.
- Global warming impacts on precipitation levels and evaporation rates, influencing flows between land and atmosphere.
- Climate change impacts on ablation rates, affecting transfers between the cryosphere and other parts of the system.

> The **cryosphere** is any place on Earth where water is frozen.
>
> **Permafrost** is ground permanently frozen for over 2 years.
>
> **Eustatic change** is the global change in the volume of water in the oceans.
>
> **Accumulation** is the build-up of snow and ice in the cryosphere.
>
> **Ablation** is the change of ice into liquid or water vapour.

3 Global systems

WJEC/Eduqas AS/A-level Geography   81

Catchment hydrology — the drainage basin as a system

REVISED

A **drainage basin** is a subsystem of the global water cycle (Figure 3.2 and Table 3.2). It is an open system because water crosses the boundaries of the basin.

> A **drainage basin** is an area of land drained by a river and its tributaries. The boundary of the basin is called the **watershed**.

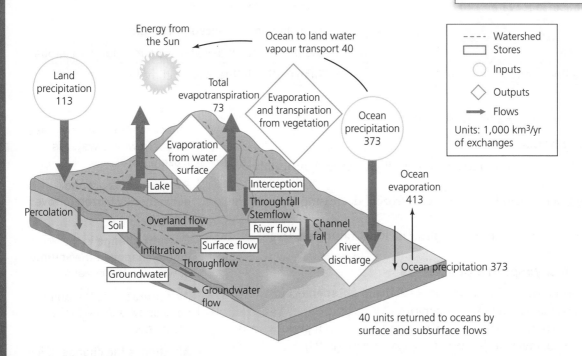

Figure 3.2 The drainage basin system

Table 3.2 The components of a drainage basin system

Component		Description
Input	Precipitation	Rain, snow, sleet, hail, frost; the type, amount, duration and intensity influence transfers and stores in the system
Flows	Throughfall	Water dripping from plants to the ground; rate is influenced by leaf cover
	Stemflow	Water flowing down stems and trunks
	Infiltration	Water soaking into the soil; infiltration rate/capacity is the rate at which water can pass into the soil; affected by soil characteristics
	Overland flow	Water flowing over the surface: ● Saturation excess overland flow — the soil is saturated, so rainfall cannot infiltrate ● Infiltration excess overland flow — rainfall intensity is so great it exceeds infiltration rate
	Throughflow	Slow, lateral (sideways) movement downslope through the soil

Component		Description
	Percolation	Downward movement from soil to underlying rock
	Groundwater flow	Slow, downward and lateral movement through bedrock
	Channel flow	Flow of water in streams and rivers
Stores	Interception store	Water held on leaf and plant surfaces
	Vegetation store	Water contained in plants
	Surface store	Water collected in depressions in the ground surface; also includes snow cover
	Channel store	Volume of water in a river channel
	Groundwater store	Water stored in underground rocks; the **water table** marks the upper level of saturated rock
Outputs	Evaporation	Water changing state from liquid to water vapour; it occurs from any surface store or flow
	Transpiration	The release and evaporation of water from vegetation
	Channel discharge to oceans	The volume of water leaving a drainage basin and flowing into the sea

Exam tip

Evaporation and transpiration are usually combined into evapotranspiration because it is difficult to measure them separately when calculating the water balance in a drainage basin.

Exam tip

It is important to use the correct terms when writing about drainage basins rather than writing generally about inputs and flows, etc.

Revision activity

Outline the factors that can influence the functioning of the different components of the drainage basin system.

Temporal variations in river discharge

REVISED

Characteristics of river regimes

- Water entering a river from groundwater creates a normal minimum flow called **baseflow**.
- Water entering by overland flow and throughflow after precipitation creates **storm flow**.
- Rivers can have a:
  - **simple regime** — one high and one low discharge corresponding to seasonal temperature and precipitation changes
  - **complex regime** — several extremes of discharge in a year

The **river regime** refers to the annual variations in a river's discharge. A line graph showing the changes is called a **hydrograph**.

Exam tip

Have examples of rivers with contrasting regimes, and be able to explain the differences.

Factors influencing river regimes

Table 3.3 Factors influencing river regimes

Physical factors	Human factors
Climate: • annual precipitation pattern • seasonal variations in temperature • evaporation rates	Land use
	Irrigation and use of water
Vegetation	Dam construction, creating reservoirs to store water and even out the flow throughout the year
Soil	
Geology	

Storm hydrographs

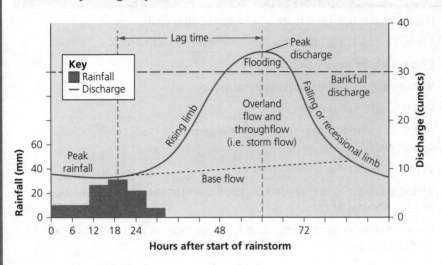

Figure 3.3 Features of a storm hydrograph

> A **storm hydrograph** shows a river's response to a precipitation event.

- Peak discharge — maximum rate of flow
- Peak rainfall — the time of the maximum rainfall
- Lag time — the period between peak rainfall and peak discharge
- Rising limb — part of the graph showing increasing discharge
- Recessional (falling) limb — shows the decreasing discharge

Factors influencing storm hydrographs

> A **flashy hydrograph** has a short lag time, high peak discharge, and steep rising and falling limbs.

Table 3.4 Factors influencing storm hydrographs

Factor		Influence
Climatic	Rainfall intensity and duration	Heavy or prolonged rain increases overland flow, resulting in a **flashy hydrograph**
	Antecedent weather conditions	Previous rainfall may saturate the ground, so overland flow occurs quickly
	Temperature	Influences evapotranspiration, affecting the amount of water available to reach the channel; frozen ground prevents infiltration, shortening lag times
River catchment characteristics	Basin size and shape	Larger drainage basins have a longer time lag; circular basins drain more quickly
		Figure 3.4 The influence of basin shape on flood hydrographs

Factor		Influence
	Drainage density	Higher densities have shorter time lags and higher peak discharge
		Figure 3.5 The influence of drainage density on flood hydrographs
	Geology and soils	Permeable rocks allow more percolation and groundwater, decreasing overland flow and discharge
		Soil type and depth influence infiltration and throughflow, affecting overland flow and lag time
	Slope angle	Steeper slopes reduce infiltration, increase overland flow and shorten lag times
	Type and amount of vegetation	Influences interception, vegetation store and transpiration rate, affecting overland flow and lag times
	Land use	Artificial, impermeable surfaces increase overland flow, reducing lag times
		Agricultural practices, such as compaction of soil by animals or machinery, or leaving land uncultivated, can influence infiltration rates
		Construction of dams and reservoirs reduces peak discharge

Revision activity

Be able to draw sketches of hydrographs to show how the shape may change as a result of changes to some of the factors listed in Table 3.4.

Now test yourself

TESTED

3 How can:
   a) geology
   b) the construction of dams
   influence river regimes?
4 What is the difference between saturation excess and infiltration excess overland flow?
5 What are the characteristics of a flashy hydrograph?

Answers on p. 171

Precipitation and excess runoff in the water cycle

Causes of air uplift, condensation and cloud formation

Three methods of air uplift result in precipitation:

- **Orographic** — air is forced to rise over higher land.
- **Frontal** — when air masses of different temperatures meet, the warmer, less dense air rises over the cooler air mass.
- **Convection** — daytime heating of the ground warms air in contact with it, forcing it to rise.

Rising air expands, resulting in cooling. When it cools to its dew point, water molecules condense around small particles (**condensation nuclei**), forming clouds.

Theories of precipitation formation

Collision coalescence process

In warm tropic areas, water condenses around large **condensation** nuclei

Larger droplets fall to Earth, colliding with smaller droplets and absorbing them

Bergeron-Findeisen process

High-altitude clouds with a temperature below 0°C contain ice crystals and **super-cooled water** droplets

The ice crystals grow as water evaporates off the droplets and deposits on the ice

The ice crystals gain weight and fall as snowflakes

Snowflakes melt in warmer air layers, forming rain

Causes of excess runoff generation

- **Prolonged precipitation:** results in saturation excess overland flow.
- **Intense storms:** result in saturation excess and infiltration excess overland flows.
- **Monsoon rainfall:** similar impact to prolonged and storm precipitation in areas of monsoon climate.
- **Snowmelt:** melting of accumulations of snow and ice results in saturation excess overland flow, especially if the ground is frozen.
- **Human causes:**
  - Urbanisation makes the surface impermeable, reducing infiltration and the soil moisture store. Drainage transports water quickly to rivers, reducing lag time. Less vegetation reduces evapotranspiration, increasing **runoff** (Figure 3.6).

Typical mistake

Do not just write about 'air going up'. Explain the process that is resulting in uplift of air. Remember that more than one process may be happening in the same place at the same time.

Revision activity

Compile annotated diagrams to illustrate the different methods of rainfall formation.

Condensation is the change from water vapour to liquid.

Super-cooled water is liquid water found below 0°C.

Runoff is water flowing over the land surface.

Typical mistake

Do not forget that precipitation may be in the form of snowfall. There could be a delay of many months between the precipitation falling and the snow melting, generating excess runoff.

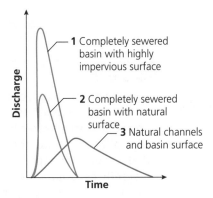

Relationship between discharge and time for three drainage basins

Figure 3.6 The impact of urbanisation on runoff

○ River basin mismanagement can increase the amount of runoff. As already seen, urbanisation increases runoff. Another example of major mismanagement is large-scale deforestation (Figure 3.7). Removal of trees increases runoff.

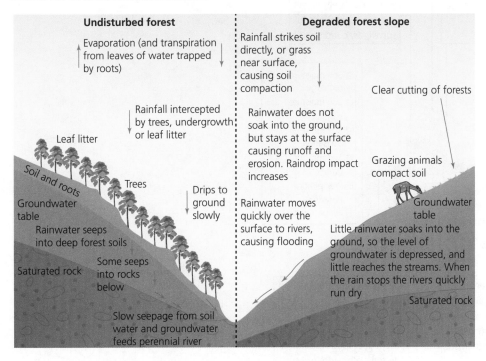

Figure 3.7 The impacts of deforestation on runoff

○ Loss of floodplain storage due to channelling and the construction of dykes for flood management results in increased runoff, creating potentially bigger floods downstream. This was a contributing factor in the River Danube floods in Romania in 2006, where 80% of floodplain storage was lost.
○ Straightening and canalisation of rivers speeds up the flow, increasing runoff downstream.

Deficit within the water cycle

Over time a water deficit will be considered a drought.

- **Meteorological drought:** an extended period of lower than average rainfall for the region.
- **Agricultural drought:** insufficient moisture to maintain crop yields.
- **Hydrological drought:** shortages in water supplies in surface and groundwater stores.

Meteorological causes

- **Seasonal precipitation variations** result in soil moisture deficit.

> **Exam tip**
>
> A water deficit will occur when evapotranspiration and runoff are greater than precipitation. Remember that it may not happen instantly because water that is stored will delay the impact.

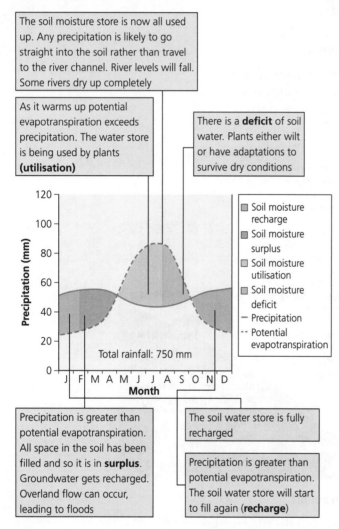

The soil moisture store is now all used up. Any precipitation is likely to go straight into the soil rather than travel to the river channel. River levels will fall. Some rivers dry up completely

As it warms up potential evapotranspiration exceeds precipitation. The water store is being used by plants **(utilisation)**

There is a **deficit** of soil water. Plants either wilt or have adaptations to survive dry conditions

Legend:
- Soil moisture recharge
- Soil moisture surplus
- Soil moisture utilisation
- Soil moisture deficit
- — Precipitation
- -- Potential evapotranspiration

Total rainfall: 750 mm

Precipitation is greater than potential evapotranspiration. All space in the soil has been filled and so it is in **surplus**. Groundwater gets recharged. Overland flow can occur, leading to floods

The soil water store is fully recharged

Precipitation is greater than potential evapotranspiration. The soil water store will start to fill again **(recharge)**

Figure 3.8 Variations in the soil moisture budget

- **El Niño Southern Oscillation (ENSO)** impacts on weather patterns around every 7 years, resulting in meteorological drought in some regions.
- **Climate change** results in long-term declining precipitation and meteorological drought.

> An **aquifer** is a layer of saturated, permeable rock.

Human causes

- **Aquifer depletion:** water is extracted from groundwater stores for human use faster than it is replenished, which can reduce water levels in rivers and lakes.
- **Surface store depletion:** the use of water reduces river discharge and the size of inland bodies of water.

> **Exam tip**
>
> Remember that the impact of aquifer depletion may not be felt until a long time after extraction begins.

Both groundwater and surface storage are used to meet increasing human demand for:
- domestic water consumption
- irrigation of crops
- industrial use

Recharge of aquifers

As well as a water deficit, over-extraction of aquifers can cause land subsidence and a long-term reduction in storage.

- **Natural recharge:** reducing extraction allows recharge by the percolation of precipitation and surface water to the aquifer.
- **Artificial recharge:**
  - Water is added to the ground surface, or pits and basins are flooded, increasing the percolation rate.
  - Injection wells — water is pumped into the aquifer via a well.

Now test yourself

TESTED

6 Why do hydrological and agricultural drought sometimes have a greater impact on humans than meteorological drought?

Answer on p. 171

The global carbon cycle

REVISED

- The global **carbon cycle** (Figure 3.9) is a closed system.
- The only inputs and outputs are between stores within the system.
- The amount of carbon is fixed, meaning that the mass balance does not change. Some carbon may be stored for millions of years.

Carbon stores (Table 3.5) are also known as **pools** or **reservoirs**. A flow between stores is also known as a **flux**. The **residence time** is the time carbon is held in a store.

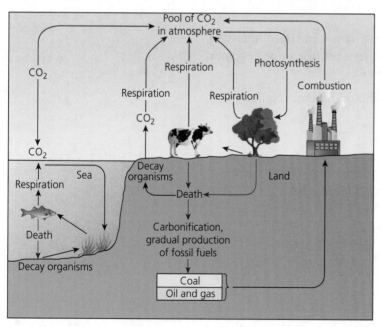

Figure 3.9 The carbon cycle

> **Exam tip**
>
> Make sure you understand why human demand for water is increasing and have examples of the deficits that have been created.

> **Revision activity**
>
> Research the location of aquifers in the UK and their importance as a store of water.

> The **carbon cycle** is a biochemical cycle in which carbon moves between the lithosphere, oceans and atmosphere.

Table 3.5 Global carbon stores

Store	Approximate percentage of global carbon	Average residence time (years)
Lithosphere and sedimentary rocks	99.9	150 million
Oceans	0.064	25–1250
Fossil fuels	0.007	150 million
Biosphere	0.003	18
Atmosphere	0.001	6

Typical mistake

While limestone and chalk are important stores of carbon, they are not the only rocks where it is stored.

Carbon pathways and processes

Between land and atmosphere at the local, short-term scale

As the residence times in the atmosphere and biosphere are relatively short, these are 'fast' carbon cycle processes.

- **Fossil fuel combustion:** burning coal, oil and gas releases carbon dioxide into the atmosphere.
- **Carbon sequestration and photosynthesis:**
  - Sequestration — carbon dioxide is removed from the atmosphere and held in solid or liquid form.
  - Photosynthesis — plants use light, water and carbon dioxide to produce glucose and oxygen.
- **Respiration:** living organisms produce energy through glucose and oxygen reacting together, resulting in carbon dioxide being released into the atmosphere.
- **Decomposition:** when organic matter dies it breaks down by physical, chemical and biological processes, which release carbon dioxide into the atmosphere.

The amount of carbon dioxide returned by respiration and decomposition is less than the amount removed by photosynthesis because some becomes part of the sedimentary rock and fossil fuel pools.

Climate can have a significant influence on photosynthesis, respiration and decomposition rates.

Between the ocean and atmosphere

Carbon dioxide moves from the atmosphere to the ocean due to carbon cycle pumps, which circulate and store carbon (Figure 3.10).

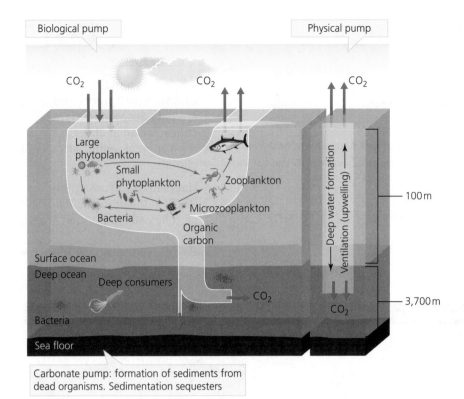

Figure 3.10 Oceanic carbon pumps

- **Absorption by biota:** phytoplankton absorbs carbon dioxide by photosynthesis, which passes through the marine food chain when the phytoplankton is eaten. Some carbon ends up in sediments when dead organisms settle on the seabed.
- **Diffusion into and out of oceans:** where the ocean is in contact with the atmosphere, carbon dioxide is absorbed by the water. Cold water sinking transfers the carbon to the deep ocean. Water moving to the surface and warming loses carbon dioxide to the atmosphere.

Between the land and oceans

The cycling of carbon between rock and oceans is a 'slow' carbon cycle because it takes a long time.

- **Weathering:** particles of rock are carried to the oceans and deposited. The chemical weathering process of carbonation dissolves the calcium carbonate in rocks, changing minerals containing lime into soluble biocarbonates that are carried by water to the oceans. There it is used by organisms to create shells, which form deposits when the organisms die, eventually creating a rock carbon pool storing carbon for millions of years.
- **River transport and indirect movement by the water cycle:** channel flow, throughflow and groundwater flow move dissolved calcium carbonate to the oceans.
- **Carbon sequestration:** deposited sediments build up over time, forming new carbon pools of sedimentary rocks.
- **Earth movements:** sediments are uplifted above sea level, where weathering processes start again.

> **Revision activity**
>
> Create a diagram to show the carbon cycle between land and the atmosphere and between land and the oceans.

Carbon stores in different biomes

Within a **biome**, carbon can be stored in:
- the biomass above and below ground
- dead organic matter on the surface (leaf litter)
- the soil

Physical conditions, such as climate, affect the **net primary productivity** of a biome, influencing the amount and location of carbon in the biome carbon pool.

> A **biome** is a major community of plants and animals adapted to the characteristics of the environment, especially climate.
>
> **Net primary productivity** (NPP) is the dry weight of biomass added per unit area per year.

Carbon stores in tropical rainforest and temperate grassland

Table 3.6 Comparison of tropical rainforest and temperate grassland carbon stores

Carbon storage	Tropical rainforest	Temperate grassland
Average NPP (kg/m^2/year)	2,200	600
World total NPP (billion tonnes/year)	37.4	5.4
Total carbon in biomass and soil (billion tonnes)	550	185
Above ground store (tonnes/hectare)	180	2–10
Below ground store (tonnes/hectare)	100	100–200
Fauna storage	High as large number of animals	Lower
Factors influencing carbon storage	**Tropical rainforest**	**Temperate grassland**
Temperature	Average annual temperature between 25°C and 30°C; minimal seasonal variation	High temperature range during the year; +40°C to –30°C possible
Precipitation	High average annual rainfall — over 2,000 mm; no dry season; convectional rain most days	Low average annual rainfall — 500 mm or less, spread throughout the year; snow in winter
Light	Sun's rays concentrated; little seasonal variation	Seasonal variation in solar radiation
Influence on the carbon stores	Year-round plant growth and high levels of biomass; atmospheric carbon sequestration is high all year. Decomposition is rapid, returning carbon quickly to the soil. High rainfall leaches the soil, removing carbon	Atmospheric carbon sequestration varies with seasonal plant growth. Humid conditions in autumn allow rapid decomposition and return of carbon to soil. Lower rainfall prevents leaching, allowing a larger soil carbon pool

Changes in the size of carbon stores due to human activity

Deforestation
- Removes biomass carbon store.
- Crops replacing trees have less biomass, reducing store size.

- Decomposition is reduced, lessening input to the soil store.
- Unprotected soil is vulnerable to erosion, reducing soil store capacity.
- Burning biomass increases carbon dioxide transfer to the atmosphere.

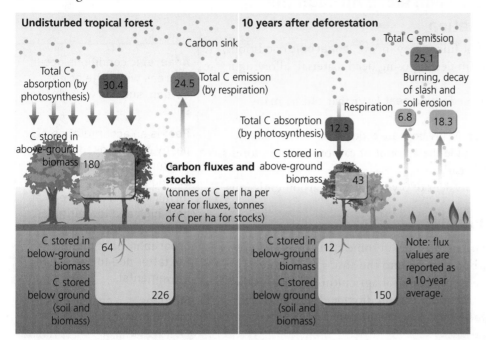

Figure 3.11 The impacts of deforestation on the carbon cycle

Afforestation

- Increases carbon sequestration and biomass store.
- Reduces soil erosion, maintaining the soil pool.

Agricultural activity

- Clearing areas for agriculture reduces carbon storage.
- Soil degradation from poor practice causes soil erosion, reducing storage capacity.
- Harvesting removes biomass store.
- Ploughing aerates soil, increasing the decomposition rate and returning carbon to the soil store.
- Over-irrigation leaches soil, removing carbon.
- Fertiliser and manure increase carbon in the soil store.
- Machinery increases carbon dioxide emissions.

> **Exam tip**
>
> Remember that afforestation only has a positive effect on carbon stores if it replaces biomass that had less storage capacity. Coniferous plantations may store less due to decreased biodiversity.

Now test yourself

TESTED ☐

7 What are the major carbon stores?
8 Why is the movement of carbon between land and oceans classified as a slow carbon cycle?
9 Why can some biomes store greater amounts of carbon than others?

Answers on p. 171

Changing carbon stores in peatlands over time

Accumulation of the carbon store through the process of peat formation

- Dead wetland plants do not fully decay in wet, **anaerobic** conditions, which prevent bacteria from decomposing dead material, allowing it to accumulate.
- Organic matter is compressed into a **peat** layer that can be many metres thick.
- Accumulation is 1 mm/year. Many bogs are thousands of years old.
- Peatlands globally store double the amount of carbon stored in forests.
- The organic matter is 50% carbon.
- UK peatlands hold the equivalent of 3 years of UK carbon emissions.

Peat extraction and drainage

- **Burning:** peat is dried and used as fuel in some rural areas and power stations. This releases carbon dioxide into the atmosphere store.
- **Drainage:** allows the use of peatlands for agriculture. Increases the decomposition rate, releasing more carbon dioxide. Increases flow of carbon to rivers. Drained areas may be afforested, changing the biomass carbon store.

Restoration of the carbon store through management of peatlands

Restoration allows peatlands to become carbon sinks, absorbing more carbon than is released. This is achieved by:
- **blocking drainage channels** — increases infiltration and raises the water table, increasing waterlogging and anaerobic conditions
- **creating berms (raised banks)** — prevents drainage
- **clearing invasive vegetation and re-establishing wetland plants**
- **adopting the 'peatland code'** — encourages investment in restoration of UK peatlands

> **Anaerobic** conditions occur in areas with little or no oxygen, such as waterlogged soils.
>
> **Peat** is an accumulation of partly decomposed organic matter, forming a deposit on boggy, acidic ground. It forms in low-relief areas at both high and low altitudes (the former due to high orographic precipitation; the latter due to soil water movements).
>
> **Methane** is a carbon-based greenhouse gas.

Links between the water and carbon cycles

Causes of recent increases in the atmospheric carbon store

The Intergovernmental Panel on Climate Change (IPCC) in 2014 stated that there was a 95–100% probability that human influence was the dominant cause of global warming from 1951 to 2010.

Carbon emissions rose from 280 ppm in 1750 to 406 ppm in 2017 due to:
- increasing fossil fuel use resulting from population growth and increasing economic growth
- deforestation (carbon dioxide)
- increased livestock farming (**methane**)
- decomposition from increasing landfill (methane)

Relationship between increases in the atmospheric carbon store and the energy budget

- 70% of solar energy reaching Earth is re-radiated back as long-wave radiation. Much of this is absorbed by the atmosphere.
- Increasing carbon dioxide and methane in the atmospheric store retains more heat, creating the **greenhouse effect** (Figure 3.12).
- Increasing temperature raises water vapour levels, which further increases temperature.

> The **energy budget** is the balance between energy received from the Sun and energy radiated back into space.

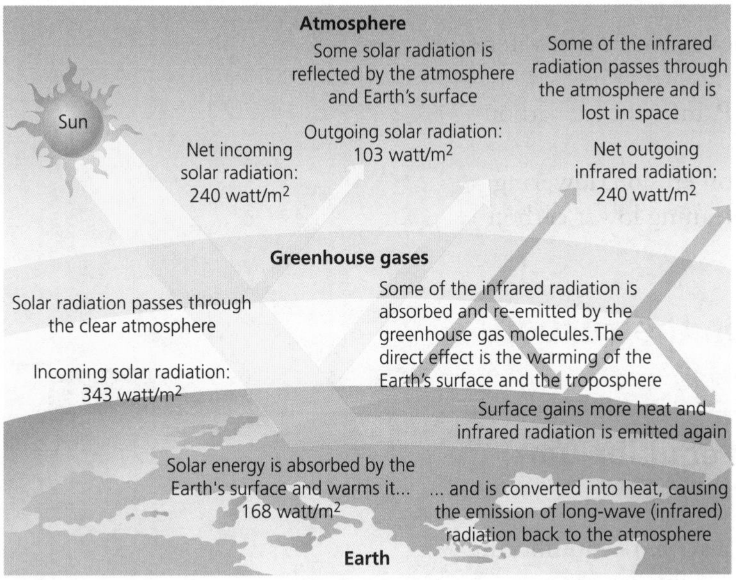

Atmosphere

Some solar radiation is reflected by the atmosphere and Earth's surface

Some of the infrared radiation passes through the atmosphere and is lost in space

Outgoing solar radiation: 103 watt/m^2

Net incoming solar radiation: 240 watt/m^2

Net outgoing infrared radiation: 240 watt/m^2

Sun

Greenhouse gases

Solar radiation passes through the clear atmosphere

Some of the infrared radiation is absorbed and re-emitted by the greenhouse gas molecules. The direct effect is the warming of the Earth's surface and the troposphere

Incoming solar radiation: 343 watt/m^2

Surface gains more heat and infrared radiation is emitted again

Solar energy is absorbed by the Earth's surface and warms it... 168 watt/m^2

... and is converted into heat, causing the emission of long-wave (infrared) radiation back to the atmosphere

Earth

Figure 3.12 The greenhouse effect

Impact of recent increases in the atmospheric carbon store on the water cycle and oceans

Precipitation and extreme weather

- Higher temperatures increase evaporation, potentially increasing precipitation.
- Increasing temperatures could lead to more heavy convectional rainfall.
- In Europe, precipitation may become more seasonal.
- Evidence suggests an increase in extreme precipitation events, heat waves and droughts. Wetter regions may experience increased precipitation.
- Hurricanes and tornados could increase in intensity.
- The IPCC has projected that global surface temperatures will rise by more than 1.5°C by the end of the twenty-first century, causing the global water cycle to change, with increasing differences between wet and dry regions.

River discharge

- Intense rainfall causes infiltration excess overland flow, rapidly increasing discharge and flooding potential.
- The river regime adjusts to precipitation changes.

Sea-level rise

- Warmer climate increases ice sheet melting, producing eustatic sea-level rise.
- Warming of the oceans expands the volume of water.
- Sea level is rising by 3.5 mm/year.

> **Exam tip**
>
> While there is general agreement on global warming, there are differing viewpoints about its impact on precipitation. While the amount of precipitation has remained the same, there has been an increase in winter heavy rain events since the 1980s. This is in line with IPCC climate change models. But in other parts of the world, notably the Sahel, there is greater uncertainty over future precipitation patterns.

Acidification of the oceans

- Increase in atmospheric carbon allows more carbon to diffuse into the oceans, creating carbonic acid. 30% increase since 1750.
- Shell-building animals have thinner shells.

Links between the water and carbon cycles at the local scale

- The cycles link where carbon is transported (dissolved or as sediment) in water.
- Changing the biomass store impacts on movement in the water cycle (e.g. deforestation).
- Irrigation alters water cycle flows and NPP, influencing carbon sequestration.
- Overgrazing and land degradation allow soil erosion, lowering infiltration rates and soil moisture store, meaning lower carbon sequestration and storage.

Feedback within and between the carbon and water cycles

REVISED

Feedback loops, thresholds and equilibrium in natural systems

Feedback is a natural response to a change in a system's equilibrium:
- **Positive feedback:** the initial change causes further change.
- **Negative feedback:** the change is reduced, restoring equilibrium.

Table 3.7 Examples of feedback

Positive feedback	Negative feedback
Increased temperatures due to enlarged atmospheric CO_2 store from fossil fuel burning	
↓	↓
More evaporation, increasing water vapour and clouds	Increase in NPP
↓	↓
More long-wave radiation absorbed	Reduced atmospheric store due to increased carbon sequestration
↓	↓
Higher temperatures	Less long-wave radiation absorbed, lowering temperatures

Without human interference most systems exist in a **steady-state equilibrium**.

Systems have a threshold — a critical level that, if exceeded, results in irreversible change. The latest evidence suggests that a global temperature rise of 1.5°C would cross the threshold in the global carbon cycle.

Exam tip
Remember links at the global scale, such as global warming, can have local impacts, such as flooding.

Steady state equilibrium is a long-term average state of balance, despite changes occurring within it.

Consequences of change within and between the water and carbon cycles

Cryosphere feedback

Table 3.8 Examples of cryosphere feedback

Positive feedback	Negative feedback
Increased temperatures ↓ Sea ice and ice caps melt ↓ Land and sea have lower albedo and absorb more heat	
↓ Atmospheric temperatures rise, melting more ice	↓ Increased evaporation and cloud cover reflect solar energy ↓ Sea temperature drops, cooling air in contact with it and decreasing melting

Terrestrial carbon feedback

Table 3.9 Examples of terrestrial carbon feedback

Positive feedback	Negative feedback
Global temperature rise	
↓ Warming tundra areas, increasing decomposition ↓ Increasing CO_2 in the atmosphere, raising temperatures	↓ Coniferous forest biome migrates northwards and expands ↓ Biomass CO_2 store expands

Marine carbon feedback

- Warmer oceans releasing more carbon dioxide contribute further to climate warming.
- Acidification impacts on the health of marine ecosystems, which reduces carbon sequestration and can impact on the biological ocean pump.

Methane feedback

- Large amounts of methane are stored in permafrost. This is released when climate change causes melting.
- Increased methane raises the atmospheric temperature, melting more permafrost.
- The impact could take Earth's climate past its threshold point.

Implications of feedback within and between the two systems for life on Earth

- Climate change impacts on vegetation patterns, influencing food production.
- Melting ice caps cause eustatic sea-level change, leading to flooding of low-lying coastal areas, where many major cities are located.
- Reduction in ice reduces annual meltwater supply, which decreases water supplies for some major cities.
- Extreme weather events cause more damage and loss of life.
- Climate change causes diseases to spread to new areas.
- Ocean temperature changes affect circulation patterns, thereby influencing climates.
- New areas become suitable for cultivation.

Exam tip

The complexity of interactions and feedback loops in the systems, combined with uncertainty about global economic, technological and political developments, mean that the implications for life on Earth are difficult to predict. Frequently, best- and worst-case scenarios are suggested.

Now test yourself

TESTED

10 How has population growth resulted in increasing carbon dioxide emissions?
11 How does the melting of ice contribute to increased global temperatures?
12 Why is it difficult to predict the consequences of changes in the water and carbon cycles?

Answers on p. 171

Exam practice

The format for the different examination papers is shown below.

Specification	Method of examination
Eduqas A-level Component 2; Section A	Two compulsory data-response questions and one extended-response question
WJEC A2 Unit 3; Section A	Two compulsory data-response questions and one extended-response question

Eduqas A-level and WJEC A2 format

1 Figure 1 shows the annual water budget for a drainage basin in northern Canada.

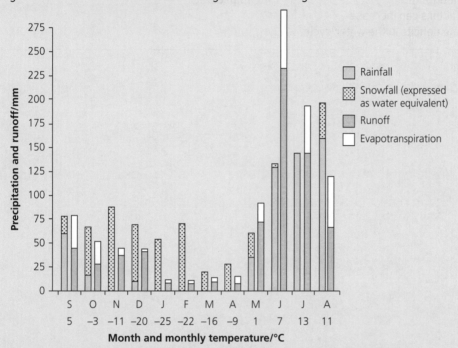

Figure 1 Annual water budget for a drainage basin in northern Canada

a) Use Figure 1 to analyse how variations in inputs can result in seasonal changes to the water budget. [5]
b) Explain how human factors can generate excess runoff in drainage basins. [5]

2 Table 1 shows carbon storage in different biomes and cropland.

Table 1 Carbon storage in selected biomes and cropland

Biome	Carbon storage (tonnes per hectare)		
	Plant	Soil	Total
Tropical forest	119	121	240
Temperate grassland	7	231	238
Cropland	2.5	80	82.5

a) Use Table 1 to analyse the variations in the pattern of carbon storage. [5]
b) Explain how human activity can influence the storage of carbon in peatlands. [5]

Eduqas A-level format

3 Evaluate the importance of different system feedback loops as a cause of accelerated global warming. Refer to both the water cycle and carbon cycle in your answer. [20]

WJEC A2 format

4 Evaluate the role of human activity in the growth of water cycle deficits. [18]

Answers and quick quiz 3 online

ONLINE

Summary

You should now have an understanding of the following:

- Both the water and carbon cycles are closed systems at the global scale, which link land, ocean and atmosphere.
- A drainage basin is an open subsystem of the water cycle.
- Long- and short-term river regimes can be influenced by natural and human factors.
- There are three methods of air uplift, creating condensation, and two ways in which precipitation may then form.
- Natural and human factors can increase overland flow or create deficits in the water cycle.
- Different parts of the carbon cycle operate at different timescales.
- Human activity has positive and negative impacts on carbon stores, including forests, grassland and peat.
- There are links and feedback between the water and carbon cycles.
- Complex changes in the systems have implications for life on Earth that are difficult to predict.

4 Global governance

Processes and patterns of global migration

Globalisation, migration and a shrinking world

REVISED

Growth of global systems, connections and global flows

Globalisation occurs in four main ways (Table 4.1).

Table 4.1 **Types of globalisation**

Type of globalisation	Examples
Economic	The growth of MNCs (p. 113) and the movements of materials, goods and investments around the globe
	The development of ICT, allowing easy global communication and data transfer
Social	International migration and multi-ethnic cities
	Increasing social interconnectivity due to technology
Political	Growth of **trading blocs** (e.g. the EU), global organisations (e.g. the World Bank) and global governance issues (e.g. use of the oceans — p. 116) and free trade
Cultural	Spread of 'Western' and increasingly non-Western culture, such as food and fashion
	Growth of **glocalisation** and hybridisation of cultures
	Increase in social media, allowing the spread of ideas

> **Typical mistake**
>
> The idea of 'global systems' is not just a modern phenomenon. People and goods were moving around parts of the globe in Roman times. The extent and speed of connections, however, have increased dramatically. This recent step change in connectivity is what most people are referring to when they talk about globalisation.

Global Governance describes the steering rules, norms, codes and regulations used to regulate human activity at an international level. 'Governance' is a broader notion of steering or piloting rather than the direct form of control associated with 'government'. On a global scale, regulation and laws can be tough to enforce.

Globalisation describes the increasing interaction between people and places at a global scale due to advances in transport and communications.

A **trading bloc** is a group of countries within a geographical region that trade together and protect themselves from imports from other countries, for example with the use of import tariffs.

Glocalisation is where products and services are distributed globally but have been adapted to meet local needs, for example McDonald's introduced the 'Chicken Maharaja Mac' to meet local tastes in India.

- The four types of globalisation are interconnected, creating global systems in which many people and goods can travel quickly over long distances.
- Globalisation results in people and places being connected to form a network.
- There are a number of global flows around these networks (Table 4.2).

Table 4.2 Types of global flow

Global flow	Description
Goods	The global demand for raw materials, food and manufactured goods has increased, especially with the growth of emerging economies such as India and China
Money	Large amounts are transacted via stock exchanges around the world MNCs based in one country invest in areas in other parts of the world
People	International migration has increased so that 3.5% of the world's population are international migrants International tourism has increased; this has been helped by tourists from emerging economies — China was the biggest spender on tourism in 2017 ($258 billion)
Technology and ideas	Fast internet speeds, social media and mobile phones allow data, media and points of view to transfer and spread rapidly

Classification of migrants

Migrants can be classified by the type of movement and the reason for movement:

- **Internal migrant:** someone who moves to a new place within the same country; this may be voluntary or forced, i.e. internally displaced people (p. 108).
- **International migrant:** someone who moves from one country to another. These can be classified as:
  - **economic migrants** — those who move voluntarily in search of work and a better quality of life
  - **refugees** — those forced to move due to fear of persecution, or death from conflict or natural hazards

Quantification and mapping of global patterns of migration

The percentage of the world's population who are international migrants has remained stable at around 3.5%. The total number of migrants has increased due to increases in the world population (Table 4.3).

> **Exam tip**
>
> For globalisation, do not just write generally about MNC fast-food chains and stores producing 'clone towns', as mentioned in Chapter 2 (p. 55). Remember to provide a structured analysis of globalisation (economic, social, political, etc.) and introduce the different types of flow that drive global systems.

Table 4.3 International migration 2017

Total number of international migrants	258 million
Proportion of working age	74%
Proportion who are women	48%
Proportion who are refugees	10%
Proportion who moved to high-income countries	64%
Proportion living in Asia or Europe	62%
Proportion of the population in Europe, North America and Oceania who are international migrants	10%

> **Typical mistake**
>
> Do not make the assumption that all migrants are poor people. Relatively wealthy people from high-income countries may also migrate for a variety of reasons.

- Asia and Europe were the regions of origin for most international migrants (Figure 4.1).
- India has the largest number of people born in the country but who live outside it.

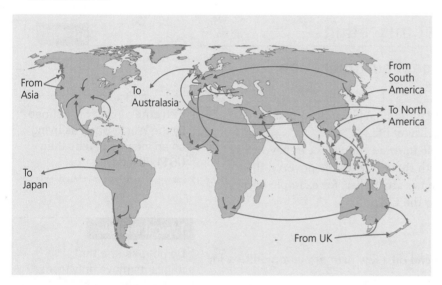

Figure 4.1 Major global migration flows in the twenty-first century

Factors creating a shrinking world for migrants

Improvements in transport and communications and increased access to media have helped increase migration.

Transport

- The travel time between places has decreased, which means distance has decreased in significance and is less of an obstacle to migration. For example, the development of high-speed railways encourages internal and international migration.
- The growth of cheap air travel allows more people to move internationally.

Communication

- The growth in mobile phone ownership and internet access allows more people to find out about opportunities in other places, encouraging migration.
- It is easier for migrants to stay in contact with family and friends in the country of origin, so there is less fear of isolation for a potential migrant.

Media

- Increasing access to media can show potential migrants the attractions of moving to a different country.
- Social media can establish groups that encourage and help migrants, for example migrants in the Netherlands are linked to Facebook groups that help with language and aid integration.
- Social media allow communication with home.

Now test yourself

TESTED ☐

1 Why can the world be referred to as 'shrinking'?
2 Why does a shrinking world potentially lead to increased migration?

Answers on pp. 171–2

Causes of international migration

REVISED ☐

Factors driving international out-migration

Poverty

- People living in **extreme poverty** and unable to afford basic needs migrate in search of a better quality of life.
- Relative poverty — while people in some countries can have an acceptable income compared with those in extreme poverty, they migrate to countries where incomes are higher, for example over 1 million Poles have migrated to the UK.

> **Extreme poverty** is defined by the World Bank as living on an income of less than US$1.90 a day.

Primary commodity prices

- Many developing countries depend on a few **primary commodities** for export revenues.
- Prices can vary by 50% in a single year due to factors such as yields or demand.
- Low prices result in a lack of investment in services such as schools and healthcare.
- Producers may not make enough to be able to afford basic needs.
- Vulnerability to fluctuating prices and the lack of investment in services can encourage people to migrate. Prices may be low due to:
  - **overproduction** — if too much is grown or yields are high, the excess supply pushes prices down
  - **poor governance** — can result in a failure to get fair trading deals as MNCs take advantage or force down prices

> **Typical mistake**
>
> Do not assume that migrants move only from low-income to high-income countries. In extreme poverty cases, the destination country may not be significantly richer than the country of origin. Many migrants move between high-income countries, for example the UK to Spain. Large volumes of 'South-South' migration also take place, e.g. movements into Nigeria (and megacity Lagos) from neighbouring countries.

Poor access to markets within global systems

- Trade blocs, such as the EU, protect their own producers by putting import tariffs on goods imported from outside the bloc. This makes it harder for countries outside the bloc to sell their produce unless the price is reduced.
- High-income countries can support their own producers with subsidies, which means that other producers have to sell at lower prices to compete.
- The lack of income from being unable to sell, or having to sell at lower prices, can encourage migration.

> A **primary commodity** is an unprocessed material that is extracted or harvested. Examples include food, minerals and energy resources.

Recent drivers of migration

International migration can also be driven by non-economic factors.

The development of diaspora communities

Migrants can be attracted to areas where people from the same country of origin already live. This can be advantageous because:

- help is given to support new arrivals
- there is the possibility of employment within the community
- there is access to services that cater for the cultural needs of the community, for example religious buildings or shops selling specific clothing or foodstuffs

India has the largest diaspora population with almost 20 million living abroad. The UK has the largest Indian community outside Asia.

> A **diaspora community** is a large group of people with a similar heritage or homeland who have migrated to different parts of the world. A diaspora includes both migrants and their descendants.

Colonial and Commonwealth links

Migrants from former colonies are attracted to the former ruling country because:

- a large diaspora community may already exist
- familiarity with the culture and language makes it easier to integrate
- they are encouraged by the former ruling nation to fill gaps in the labour force, for example the 'Windrush Generation', who were encouraged after World War II to move from the Caribbean to live and work in the UK

> **Revision activity**
>
> Research links that you may have to a diaspora community. Are you part of one or do you know someone who is? Is there one close to where you live? How can you tell when you are entering an area where a diaspora community exists?

Legislation permitting freedom of movement

- Parts of the world allow free movement of migrants between countries.
- The EU allows free movement of labour. Large numbers have migrated from Eastern Europe to the UK, France and Germany.
- The Mercosur Residence Agreement allows nationals of nine South American countries to live and work in other member countries.

> **Revision activity**
>
> Research the Schengen Agreement and why Britain decided to opt out of it.

The influence of superpowers

The influence exerted by **superpowers** on other countries can make them an attractive destination for migrants:

- The USA's involvement in world affairs since World War II includes:
  - international aid
  - military power
  - economic influence, such as economic sanctions
  - the spread of American culture and media
- Many European countries, such as the UK, France, Spain and Germany, had empires with colonies in different continents, often spreading European languages and culture. Familiarity results in migrants from former colonies moving to the former governing country.
- China is now the world's largest economy, with its foreign direct investment (FDI) having an increasing global impact.

The perceived opportunities of a better quality of life, familiarity with the language and culture, as well as an existing migrant community, makes many superpowers an attractive destination for migrants (Table 4.4).

> A **superpower** is a country that can exert influence and power (consisting of economic and military 'hard power' along with cultural and/or diplomatic 'soft power') at a global scale. A country that can exert such an influence over its neighbours but not globally is a regional superpower.

Table 4.4 Number of immigrants resident in G7 countries in 2017

G7 country	Total number of immigrants resident in the country in 2017 (millions)	Total number of immigrants as a percentage of total population
USA	49.8	14.5%
Germany	12.2	14.9%
UK	8.8	13.2%
Canada	7.9	21.9%
France	7.9	12.1%
Italy	5.9	9.7%
Japan	2.2	1.7%

> The **G7** is a group consisting of seven of the most powerful countries in the world.

- Some regional superpowers are attractive to migrants. The migrant population of the Middle East countries doubled to 54 million between 2005 and 2015.

The advantages of international migration for superpowers

- Migrants fill labour shortages.
- More skilled workers, for example Indian doctors working in the NHS.
- Many migrants are willing to work in unskilled and/or low-paid jobs such as cleaning or restaurant work.
- MNCs are often keen to invest in areas with a diverse population and may employ highly skilled migrants, for example the London-based Deutsche Bank's former CEO was born and raised in India.

> **Typical mistake**
>
> Do not assume that all superpowers actively encourage immigration. Japan has strict residency rules, which combined with the difficulties of the language for many, means that migration figures are much lower than for the other G7 countries.

The development of cities as global hubs

- Cities may be considered as **global hubs** due to the presence of MNC headquarters, globally renowned universities or global financial or political institutions.
- Many global hubs have benefited from some form of 'top-down' support from the government.
- China has created planned industrial hubs, for example Shenzhen was made a special economic zone, attracting the headquarters of many Chinese high-tech companies, MNCs and financial institutions.
- Silicon Valley in California had US government support, such as defence contracts for research and development, and work for government departments.
- Other strategies include grants and tax breaks, public–private partnerships and building infrastructure to attract investment in the city.

> A **global hub** is a city or region that provides a focal point for activities at a global scale.

- Global hubs usually continue to grow over time, aided by national and also international migration (Figure 4.2), for example staff working for British MNCs have migrated overseas to help manage their companies' Chinese and US offices.

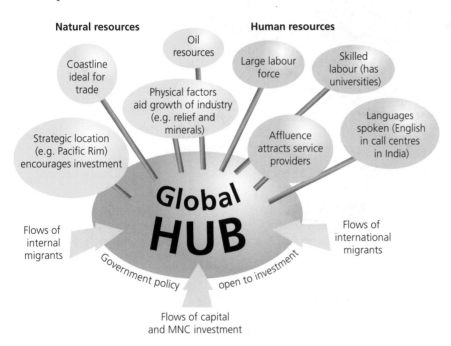

Figure 4.2 The role of migration in the growth of global hubs

The consequences and management of international economic migration

REVISED

Flows of money, ideas and technology linked with economic migration

- Large numbers of low-skilled migrant workers result in money leaving the host country as **remittances** to the migrants' home nation (Figure 4.3).

> **Remittances** are funds transferred by foreign workers to their family and friends in their home country.

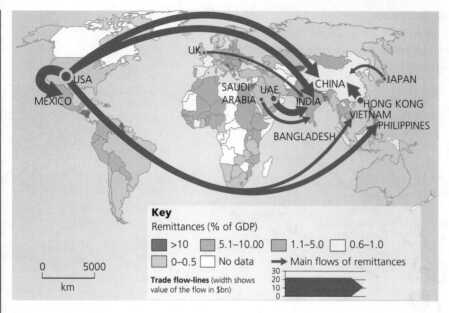

Figure 4.3 Flows of migrant remittances 2011

- ○ Remittances can help reduce poverty and increase spending in migrants' country of origin, creating a positive multiplier effect and possibly reducing inequalities.
- ○ India received the most in remittances in 2017, with a value of $69 billion.
- ○ Remittances are worth between two and three times the amount of the global development aid budget.
- ○ In nine developing countries, remittances contribute over 20% of GDP.
- ● Skilled migrant workers can create a 'brain drain' in the source nation as they leave to work abroad, for example Indian professionals working in high-tech industries in the USA.
- ● Flows of money, ideas and technology linked with economic migration can promote growth and stability, but also exacerbate inequality, causing conflict (Table 4.5).

Table 4.5 Consequences of economic migration

	Host nation	Source nation
Promotion of growth and stability	Labour skill shortages can be filled, which may increase productivity Labour force willing to accept lower-paid jobs that locals do not want Many migrants have entrepreneurial skills and establish successful new businesses, providing employment opportunities, for example a seventh of UK companies with a turnover of over £1 million a year were founded by migrants Positive multiplier effect of migrant spending can help boost the economy	Spending remittances in the source nation boosts the economy, for example money sent back to the Philippines after the global financial crisis of 2008 prevented the country going into recession Migrants and descendants may return home and establish businesses using skills learned in the host nation

	Host nation	Source nation
Exacerbation of inequalities	Skilled migrants can bring new ideas and skills to companies in the host nation, which can increase inequalities with source nations	The loss of a large number of the young, economically active population results in a lack of investment in economic activities
	The influx of migrants can increase pressure on services such as schools, creating a financial burden for areas where they settle; this may create conflicts with locals unwilling to contribute to the costs	The lack of spending by workers slows economic growth, increasing inequalities with host nations
	People in the host nation may have the impression that migrants are taking jobs from locals, which may create conflict	Economic losses result from the costs of educating and training workers who then go abroad
	Migrants may fail to integrate into the host country, creating social conflicts	Political changes may see migrants forced to leave host nations and return home, creating conflicts between the countries

Economic, social, political and environmental interdependency

- Globalisation has helped create an increasing level of **interdependency** between countries.
- The growth of global trade and methods of production where components are manufactured and assembled in different countries has led to the growth of economic interdependency, for example a Land Rover Discovery made in the UK uses parts from the UK as well as from Germany, France, Hungary, Ireland, Czech Republic and Poland.
- A family relying on a member finding work abroad and sending remittances, so that the family can improve its quality of life, demonstrates an example of social interdependency.
- Countries joining together in intergovernmental organisations such as the World Bank, United Nations (UN) and the EU allows member states to assist each other in matters such as financial issues, trade and conflict resolutions, illustrating political interdependency.
- Environmental interdependency can be seen when environmental concerns have resulted in countries working together to find solutions, such as the 1997 Kyoto Protocol to control greenhouse gas emissions.
- Interdependency can create benefits and risks to countries (Table 4.6).

> **Exam tip**
>
> Remember that migration can have consequences for both host and source nations. Check carefully if you are being asked to write about both.

> **Interdependency** is when two or more countries are mutually reliant on one another.

Table 4.6 The benefits and risks of interdependency

Benefits	Risks
• Decrease in conflict — countries relying on each other may be less likely to go to war with each other • Countries joined in intergovernmental organisations can work together to solve problems, for example the World Bank gave US$1.6 billion to support countries hardest hit by the spread of Ebola in 2015 • In organisations such as the EU, migrant labour allows economic output to be maximised and the greater revenues are shared between all members	• An economic recession can result in cancellation of construction projects, which may result in unemployment for migrant labour and fewer remittances being sent home • Events in the source nation can result in migrant workers returning home, creating labour shortages in the host nation, for example Australians returning home for the Olympics in 2000 created difficulties in finding temporary teachers to fill vacancies in parts of the UK • The process of **backwash** results in migrants moving from peripheral regions to hubs, which leaves labour shortages in some areas • Increased resentment against migrants as people feel the migrants are taking the available jobs

Migration policies of host and source countries

Migration policies vary between countries and reflect the economic, social and political circumstances of each country at the time (Table 4.7).

Table 4.7 Migration policies adopted by countries 2017

Policy	Percentage of countries adopting policy in 2017
Maintain the current level of migration	61
Reduce the level of migration	13
Increase the level of migration	12
No official policy	14
Increase the proportion of highly skilled workers	44
Encourage the return of their citizens	72

> **Exam tip**
>
> The issue of migration is frequently in the news. Stay up to date with events to keep your answers relevant.

The UN's 2030 Agenda for Sustainable Development called on countries to facilitate orderly, safe, regular and responsible migration, through the implementation of planned and well-managed migration policies.

Of the 20 countries with the largest number of citizens living abroad, 15 have a policy to encourage their citizens to return, including Mexico, Poland and China. India, Bangladesh and Germany do not have such a policy.

Table 4.8 outlines part of the migration policy adopted by the UK as a host country. The UK has some of the most complex migration rules in the world.

Table 4.8 The UK's migration policy for non-EU nationals 2018

The UK's five-tier, points-based system for migration	
Tier 1	Entrepreneurs, investors and exceptional talent After 3 years applicants can apply to remain if they have access to £200,000
Tier 2	Skilled workers Points are awarded for qualifications, English language skills and future expected earnings There is an annual quota of 20,700 people
Tier 3	Unskilled workers Not available due to migrants from the EU meeting the UK's needs
Tier 4	Students coming to study Must be able to speak and write English Must be able to afford to support themselves and pay course fees
Tier 5	Short-term temporary workers, such as entertainers, athletes and charity workers

Exam tip

Remember that Table 4.8 is a very simplified summary of some complex migration policies. These policies can change suddenly, for example in 2018 the UK government announced an increase in the number of skilled workers allowed to enter in order to fill doctor shortages in the NHS. Try to stay up to date with changes.

Conflicting views about cultural change and migration

Concerns over the impacts of globalisation and migration have resulted in differing views on the subject.

- In the 2016 referendum on UK membership of the EU, immigration was one of the main issues influencing the vote.
- In 2017 a new poll suggested that 52% of UK adults believed more migrants should be admitted to do highly skilled jobs. 40% said that fewer lower-skilled migrants should be admitted.
- Areas with high numbers of migrants, such as London, are more willing to accept migration than areas less influenced by migration.
- Migration and a fear of losing a national identity have led to an increase in support for nationalist parties in some EU countries.
- While a diaspora community can influence the culture, such as the food, music and language of an area, and can cause conflicting views among people, it can be advantageous. For example, Birmingham attracts people by advertising the 'Balti Triangle' as a centre for Asian cuisine and fashion.
- A person's views on immigration tend to be influenced by their personal experience of it, which is often influenced by where they live.

Now test yourself

TESTED ☐

6 How can remittances benefit a source nation?
7 Why can interdependency bring risks to a country?

Answers on p. 172

Causes, consequences and management of refugee movements

- **Refugees** can be forced to leave due to human factors, such as geopolitical reasons (e.g. conflicts, persecution or economic actions). They can also be forced out by natural disasters, such as drought.
- The total number of forcibly displaced people (refugees, IDPs and asylum seekers) rose from 33.9 million to 65.6 million between 1997 and 2017 (Table 4.9). Most of the increase occurred between 2012 and 2015.

Refugees are people who have been forced to leave their home country. **(Internally displaced people** (IDPs) have been forced to leave their homes, but remain within their own country.)

Table 4.9 Forcibly displaced people movements 2017

Type of displacement	Number of people (millions)
Total forcibly displaced	65.6
Refugees	17.2
IDPs	40.3
Asylum seekers	2.8
Main source nations	
Syria	5.5
Afghanistan	2.5
South Sudan	1.4
Top five host nations	
Turkey	2.9
Pakistan	1.4
Lebanon	1.0
Iran	0.97
Uganda	0.94

Typical mistake

Do not assume that refugees are all unskilled. Many refugees are well educated and highly skilled. Some may have run successful businesses.

Causes of refugee movements and internal displacement

Geopolitical causes

- As a result of their colonial past, many countries in Africa and the Middle East have borders that bear little relation to ethnic groups, and this has resulted in conflicts over power.
- Powerful countries such as Russia, the USA, the UK and France, each with their own agenda, have become involved in supporting different sides, prolonging conflict.
- People flee from conflict and war, for example a multi-sided armed conflict in Syria began in 2011 with factions fighting against the president and each other. By April 2018, 5.6 million Syrians had become refugees, along with 6 million IDPs.
- People flee from persecution. This often occurs during, and as a result of, war.
- As well as political factors, there may be religious, ethnic, racial or national persecution. For example, in the Central African Republic in 2013 the Muslim 'Séléka' group overthrew the Christian president and many Christians were persecuted, who in turn retaliated and persecuted Muslims. The ongoing conflict has resulted in millions of displaced people.

Revision activity

Research a recent geopolitical event that has resulted in refugee movements. Where did people move from and where did they move to? How many people became refugees? If you are using the examples given here, make sure the figures are up to date when you come to do the examination.

Economic injustice causes — land grabs

- Land grabbing occurs when individuals lose access to land they previously used, threatening their livelihood.
- Large areas of land in low-income countries are acquired by MNCs, foreign governments and individuals.
- Companies acquire land to make money from cash crops for export, such as palm oil, soy, sugar cane and biofuels.
- Around 60% of Cambodia's arable land has been handed to private companies, displacing many people.
- Africa is a prime target for land grabs — 70% of land acquired by foreign investors is in just 11 countries, seven of which are in Africa (e.g. Ethiopia and Sudan).

Natural disasters and climate change

- Natural hazards leading to disasters can cause people to flee to safety (p. 155).
- Over 20 million people in Somalia, South Sudan, Nigeria and Yemen are facing extreme drought and are moving to avoid starvation.
- In the future, climate change could result in 13 million coastal dwellers being displaced as a result of rising sea levels (p. 95).

The consequences of refugee movements

Impact on the lives of refugees

- Refugees flee homes, leaving possessions, jobs, social networks and sometimes family members, often risking their lives to reach the destination.
- A lack of money, citizenship and the possibility of not speaking the language can make integration in the host region difficult.
- Many refugees are held in refugee camps where conditions are often overcrowded and basic.
- Refugees are unable to work and rely on aid organisations for many basic needs. Children may receive no education.
- In the UK a refugee given permission to stay has 28 days to find accommodation and apply for benefits before they are evicted from asylum accommodation. Many then become homeless.

Impact on neighbouring states

- Most refugees travel the shortest distance until they feel safe. This can result in countries surrounding an area where people are fleeing receiving large numbers of refugees. For example, countries surrounding Syria have received far more Syrian refugees than EU countries further away (Figure 4.4).

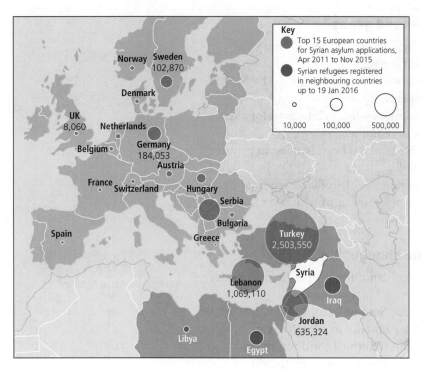

Figure 4.4 Destination of Syrian refugees 2016

- Large numbers can put pressure on surrounding areas to provide refugee camps, humanitarian aid and support for refugees.
- Refugees may never return home, integrating into the host country. For example, the Dadaab refugee complex in Kenya was established in 1991 for refugees from the civil war across the border in Somalia. More refugees arrived after fleeing drought in 2011. The early part of the complex is like a small town, with commercial hubs linking the two countries. Children and grandchildren have been born and set up home there.

Impact on developed economies

Countries agreeing to the Universal Declaration of Human Rights are obliged to provide refugees the 'right to seek and enjoy asylum'.

Disadvantages:
- The cost of supporting a refugee can be high — over £15,000 per refugee in the first year of relocation. The UK government Department for Work and Pensions estimates costs of £12,700 in benefits. Local authorities estimate an additional cost of £8,520, which covers housing and other services. Medical costs can add another £2,000.
- Refugees tend to assimilate slowly into the host countries, often because they have been traumatised and forced to leave. They are often less well-matched to a host country than migrants who choose to travel there.
- Refugees are often mistaken for illegal immigrants or economic migrants. In some European countries it is thought by some people that the number is too high and this has influenced political views.

Benefits:
- Refugees can fill labour shortages or create jobs.
- Former Ugandan Asian refugees have created an estimated 30,000 jobs in Leicester since their arrival in 1972.

Revision activity

Complete an internet search to find images of the Dadaab refugee complex. Use your findings to assess the quality of life for the refugees. What impact do you think the complex has on Kenya?

- Refugees may have skills that can be used, saving the costs of training new workers. For example, 1,200 medically qualified refugees work in the UK. It costs £25,000 to support a refugee doctor to work in the UK but costs £250,000 to train a new one.

Actions to tackle refugee crises

Governments and NGOs have policies to deal with refugee crises.

National governments

- The policy on refugees adopted by national governments varies greatly, resulting in a large variation in the number of refugee applications for asylum being accepted (Table 4.10).

Table 4.10 Accepted asylum applications for selected countries 2016

Country	Accepted asylum applications
Germany	433,905
USA	84,989
Sweden	66,585
Italy	35,405
France	28,755
Australia	25,750
UK	9,935
Greece	2,715
Ireland	485

- In 2016 the EU accepted 45% of all asylum applications. The figures for individual member countries ranged from 94% in Bulgaria to 9% in Hungary. The UK accepted 39%.
- Of the 25,750 asylum applications accepted by Australia in 2016, 12,000 were from Syria and Iran.
- Military vessels patrol Australian waters and tow asylum seekers back to Indonesia.
- Refugees reaching Australia are held in an offshore processing centre on the Pacific island of Nauru. If accepted, they are settled in Nauru.

Office of the United Nations High Commissioner for Refugees (UNHCR)

This is an intergovernmental institution supported by many countries, whose role is to provide international protection and seek permanent solutions to the problem of refugees. UNHCR services include:
- emergency assistance, such as clean water, sanitation, healthcare, shelter, blankets and household goods
- transport and assistance for refugees returning home
- training and income-generating projects for refugees who resettle
- cash-based interventions — cash and vouchers are given to refugees so that they can buy food and access services

During 2018, the UNHCR proposed the 'Global compact on refugees' to the UN General Assembly. It aims to ease pressure on host countries, enhance refugee self-reliance and support conditions in source countries so refugees can return safely.

The work of NGOs

NGOs aim to ease refugee crises in a variety of ways, such as providing aid and support to refugees as well as advice to help governments. Many charities devote some of their work to helping refugees. Examples include:

- Refugee Action — provides support and advice for refugees settling in the UK.
- Refugees International — a global, independent organisation that challenges governments and policy makers to improve the lives of refugees; it carries out field research and reports back to policy makers.

The powerlessness of some states in relation to cross-border flows of people

Some states are relatively powerless to prevent the movement of people or resources across their borders:

- A country may not have the resources to be able to protect and control its borders.
- The borders may be long and pass through relatively remote areas that lack infrastructure and government services, making it difficult to identify the actual border.

An example is the Democratic Republic of the Congo (DRC), which is bordered by nine countries. For 15 years up to 2010, militia groups entered the DRC to support ethnic groups. In fear, many refugees crossed uncontrolled borders into the Central African Republic. The militia groups have also decimated DRC's wildlife, illegally trafficking wildlife products out of the country to South Sudan, Uganda and Kenya.

Now test yourself

TESTED

8 What is the difference between a refugee and an economic migrant?
9 What benefits can refugees bring to a country?

Answers on p. 172

Rural–urban migration in developing countries

REVISED

- Since 2010 an estimated 3 million people a week have been moving to cities.
- By 2030 almost 10% of Earth's land surface will be urbanised.
- The greatest rural–urban movement is in developing countries, with a growth in **megacities**.

> A **megacity** is a city with a population of 10 million or more.

Push factors in rural areas

- Poverty in rural areas is the main push factor encouraging people to leave in search of a better quality of life.
- Farming can involve physical and demanding work, which many may not want to do.
- Rural areas can lack basic services and education can be limited.
- People may be displaced from the land in rural areas as a result of land grabs or fleeing from persecution (p. 108).

Mechanised agriculture

Mechanisation reduces the need for labour, for example the introduction of tractors to Burkina Faso eliminated in the need for male labourers tilling the land.

The role of MNCs

- MNCs are involved in cultivating and exporting cash crops grown in developing countries. These can be large-scale operations, for example Karuturi Global Ltd, the largest producer of cut roses in the world, leased 25,000 ha of land in Ethiopia in 2018 to grow flowers for export.
- In developing agriculture, MNCs may introduce modern technology to improve productivity, thus reducing job opportunities.
- MNC involvement in agriculture can result in land grabs, which prevent indigenous peoples accessing their land (see below).
- Jobs created by MNCs can be low paid and lack security, for example Karuturi Global has previously abandoned a number of large commercial farms, leaving workers unemployed.
- MNCs may develop infrastructure, such as roads, to help distribute the cash crops. This can make it easier for migrants to travel to urban areas.

Displacement of indigenous peoples by global systems

Land grabs by MNCs can result in individuals losing access to land they previously farmed. These people may migrate to cities in search of a new livelihood (see below).

Land degradation

Over 3.6 billion hectares of arid land are affected by desertification and 10 million hectares deteriorate each year. Reduced yields and abandoned land result in increased poverty, which encourages migration.

Employment pull factors in urban areas

The employment opportunities in urban areas in developing and emerging economies are the dominant pull factor in rural–urban migration. These opportunities can increase as a result of a number of factors.

Global supply chains

- MNCs have moved factories and offices abroad to take advantage of lower wage costs. This process of **offshoring** creates employment opportunities in countries such as China, India and those in South America and Africa.
- MNCs also **outsource** production, using foreign companies to produce goods or provide services, for example Apple outsourcing to China. Nike outsources footwear production to Thailand, South Korea, Vietnam and India.
- The growth in demand for information technology (IT) services has led to the growth of outsourcing and offshoring of IT services, especially to India by companies such as HP, IBM and Microsoft. These jobs are more attractive than working in agriculture, especially for young, educated people.

Typical mistake

When discussing push factors do not just talk vaguely about, for example, 'lack of jobs'. Use specific examples of why people migrate from rural to urban areas.

Exam tip

If asked to write about rural–urban migration, do not just mention poverty as a push factor. Explain the reasons for the poverty that encourages people to leave. Use examples to add detail to your answers.

A **global supply chain** is the worldwide network an MNC relies on to produce and distribute its goods and services.

Revision activity

Research examples of MNCs that have taken part in offshoring. Where have they offshored to? What part of their operation is offshored? Why do you think this has happened?

Export processing zones

- The development of EPZs has encouraged MNCs to invest in many developing and emerging economies in the world.
- EPZs have increased by over 3,000 in the last 20 years, so that there are now over 4,000, creating over 70 million jobs.
- Manufacturing in many EPZs in developing countries provides jobs for young, unmarried, poorly educated females whose employment opportunities are often limited in rural areas.

Other opportunities

A large urban area may offer informal job opportunities for migrants, such as recycling materials found at landfill sites in cities like Nairobi.

> **Export processing zones** (EPZs) are industrial areas especially set up by governments to attract foreign investment and create employment. Companies operating in EPZs may get incentives, such as duty-free imports of raw materials, flexibility in labour laws and tax concessions.

Consequences of rural–urban migration

Rural areas

- Young, economically active and ambitious people, especially men, are most likely to migrate.
- Elderly people who remain may be the least able to do the physical agricultural work. For example, 70% of India's population aged over 60 are rural dwellers.
- The most educated often leave, so rural areas become less likely to be introduced to new, innovative ideas.
- Changes in the workforce make rural areas less productive. This could make the areas more vulnerable to land grabs.
- Migrants can send remittances back to family members in rural areas, which support the population.

In 2013, 80% of 193 UN countries had policies to lower rural–urban migration. Most of these policies were proving ineffective. Possible solutions to the problem include:

- countries investing in agriculture to make it a more attractive occupation, for example investment in the production of cocoa in Ghana led to the return of 2 million Ghanaians
- investing in rural areas so that people are less likely to leave
- improving services such as healthcare and education, for example free healthcare and housing improvements reduced migration in Sri Lanka
- improving infrastructure, such as water supply and sanitation, as well as transport links to aid the transport of agricultural produce to market or processing factories

Some policies can have unforeseen consequences. The attempt to develop more rural industries in India increased migration, as rural dwellers found the new skills they had attained were in demand in the cities, where they could earn more money.

Urban areas

- Rural–urban migration results in the rapid growth of cities and an increase in the number of megacities with a population of over 10 million (Figure 4.5). For example, Lagos in Nigeria is expected to increase from 10 million to 34 million people by 2050.
- It is estimated that 3 million move to cities globally every week, a large proportion of them from rural areas.

- As well as larger populations, cities also increase in size, which can add to problems of congestion and air pollution. For example, Nairobi in Kenya is expected to increase to six times its present size by 2050.

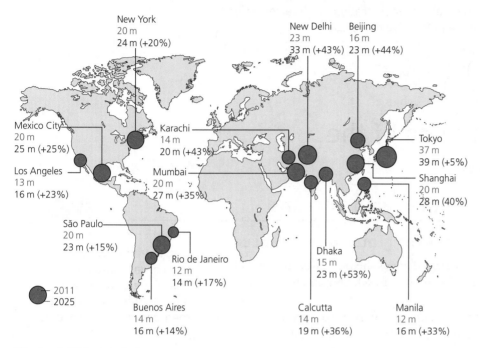

Figure 4.5 The projected growth of megacities 2011–2025

- A lack of suitable housing and the inability to meet the demand for housing in cities lead to the growth of squatter settlements on the edges of urban areas and in other parts of the city not suited for housing.
- Jobs available to migrants in EPZs are often poorly paid, with poor working conditions due to relaxed labour laws. The workers have little money so cannot invest in improving living conditions.
- The cost of installing infrastructure in settled areas can be nine times higher than installing it before development.
- Slum areas can become important to the economy of the city. Dharavi in Mumbai has over 7,000 businesses and 15,000 single-room factories, with a turnover equivalent to £700 million a year.
- Managing rapid urban growth in developing megacities can involve top-down methods, where authorities plan and provide for the growing population, for example the construction of housing developments and new cities in China.
- Bottom-up models involve people working together to make their own improvements. The organisation Practical Action supports community-led schemes in squatter settlements around the world to improve water supply, waste disposal and sanitation.
- Management methods may be a combination of top-down and bottom-up schemes. For example, in Rocinha in Rio de Janeiro, the authorities supported bottom-up schemes by providing residents with materials for building permanent accommodation. The money saved on labour allowed top-down plans to provide services such as water, electricity, schools and health centres. The area is now home to many businesses.

Exam tip

At A-level you need to write in detail about the different impacts in urban areas and to support your answers with examples. You cannot just write about rural–urban migration leading to the growth of shanty towns.

Now test yourself

TESTED

10 What is the difference between outsourcing and offshoring?
11 Why do EPZs contribute to rural–urban migration in some countries?
12 What is the difference between top-down and bottom-up developments?

Answers on p. 172

Global governance of the Earth's oceans

Global governance of the Earth's oceans

REVISED

Supranational institutions for global governance

- Governance of the Earth's oceans includes the processes, agreements and rules developed to organise the way in which humans use the oceans and their resources.
- As the oceans are beyond the territorial waters of countries, agreements are made by **supranational institutions** (Table 4.11).

> A **supranational institution** is an organisation that shares decision making and has influence over its member states, regardless of national boundaries.

Table 4.11 Supranational institutions with a role in the global governance of oceans

Supranational institution	Structure/role
United Nations (UN)	Founded in 1945, it now has 193 members. Its role includes: • maintaining peace and security and upholding international law • protecting human rights • providing humanitarian aid • promoting sustainable development The fulfilling of some of these roles will involve ocean governance.
United Nations Educational, Scientific and Cultural Organization (UNESCO)	Part of its role is to protect the environment, including oceans. Its Intergovernmental Oceanographic Commission (IOC) promotes cooperation and coordinates marine research in effectively managing ocean resources.
EU	A political and economic union of European states. Its Marine Strategy Framework Directive aims to protect the marine environment in Europe.
G8/G7	Intergovernmental political forum made up of the USA, the UK, Germany, Italy, France, Japan, Canada and Russia. The exclusion of Russia in 2014 created the G7. Members meet at least annually to discuss economic policies. In 2017 the G7 made the future of the oceans a key priority, aiming to conserve and sustainably use the ocean resources.
G20	International forum of 19 countries and the EU. It replaces the G8 as the main economic forum of wealthy nations. It has called for global ocean governance to ensure that the use of oceans is sustainable. In 2017 it developed an action plan to reduce plastic and waste entering oceans.

Exam practice answers and quick quizzes at **www.hoddereducation.co.uk/myrevisionnotesdownloads**

Supranational institution	Structure/role
G77	Founded in 1964 as a coalition of developing countries to promote economic interests. It now has 134 members. Committed to supporting the UN's goal of conservation and sustainable use of the oceans.
North Atlantic Treaty Organization (NATO)	An intergovernmental military alliance between 29 North American and European countries. Plays a role in maritime security not just in the Atlantic, for example in tackling piracy in the Indian Ocean.

Laws and agreements regulating the use of the Earth's oceans

There are a number of laws and agreements relating to the governance of the oceans that are designed to aid geopolitical stability and the protection and sustainable development of the oceans.

United Nations Convention on the Law of the Sea (UNCLOS)

- This defined the rights and jurisdiction countries have over different parts of the ocean by establishing territorial sea limits (Figure 4.6).
- The convention came into force on 16 November 1994 and has been signed by 168 countries.
- Some landlocked countries and the USA have not signed.

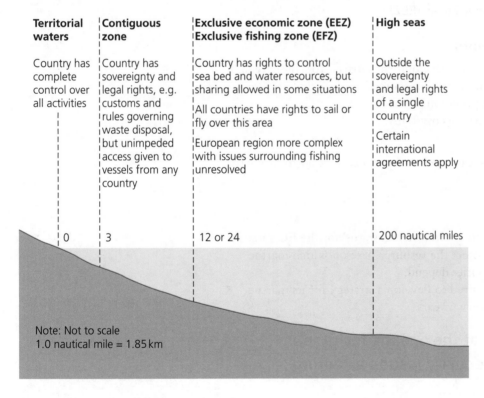

Figure 4.6 The UNCLOS ocean management zones

- The creation of Exclusive Economic Zones (EEZs) gave coastal countries ownership of resources in the water and ocean floor up to 200 nautical miles (370 km) from the shore (Figure 4.7).
- Where two countries' EEZs overlapped a decision was made on the boundaries.

- This led to disputes, for example Norway and Russia dispute the area around Svalbard.
- Countries also claimed EEZs for overseas territories, for example the UK and Argentina are in dispute over the EEZ around the Falkland Islands.

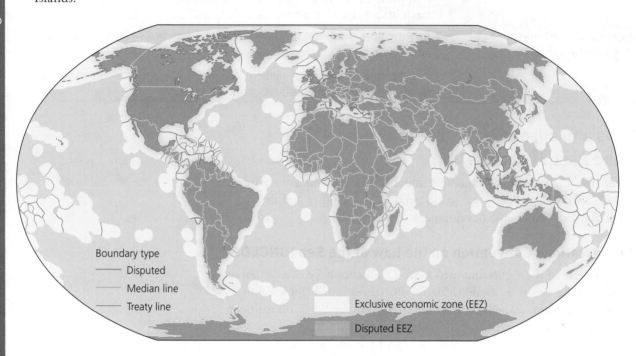

Boundary type
— Disputed
— Median line
— Treaty line

Exclusive economic zone (EEZ)

Disputed EEZ

Figure 4.7 Exclusive Economic Zones (EEZ)

Sustainability agreements

Goal 14 of the UN's Sustainable Development Goals is to 'conserve and sustainably use the oceans and marine resources'. This includes:
- preventing and reducing pollution
- sustainably managing marine ecosystems
- ending overfishing

In 2017 the UN started a 2-year process to produce a treaty to protect wildlife in the high seas.

EU Marine Directive

- This aims to achieve 'good environmental status' for the EU's marine waters by 2020 and to protect the resource base on which marine economic and social activities depend.
- Each member state is required to develop a strategy for its marine waters that is renewed every 6 years.

The strategic value of the oceans for global superpowers and security issues affecting maritime trade

Controlling the oceans allows superpowers to extend their sphere of influence in the world. This may be through involvement in:
- protecting sea lines of communication and trade, for example by upholding international law
- ensuring maritime security
- preventing conflict
- delivering humanitarian operations and emergency relief

The USA and China are two of today's marine superpowers, while India is emerging as a new superpower. Russia is trying to regain its superpower position.

The strategic value of the oceans for resources, and global trade and communications, has created a number of potential security issues that could have a serious impact on superpowers and other nations.

Oil transit chokepoints

- Over 66% of the world's oil is moved by sea. Around 53% of oil travels through seven major **chokepoints** (Table 4.12).

> **Chokepoints** are narrow channels along widely used global sea routes.

Table 4.12 Oil chokepoints

Chokepoint	Location	Amount of oil transported through chokepoint in 2016 (million barrels per day)
Strait of Hormuz	Between the Persian Gulf, the Gulf of Oman and the open sea	18.5
Strait of Malacca	Between the Indian and Pacific Oceans	16.0
Suez Canal	Between the Red and Mediterranean Seas	5.5
Mandeb Strait	Between the Red Sea and Indian Ocean	4.8
Danish Strait	Between the North and Baltic Seas	3.2
Turkish Strait	Between the Aegean and Mediterranean Seas	2.4
Panama Canal	Between the Atlantic and Pacific Oceans	0.9

- Blocking a chokepoint can have a number of impacts:
  - increasing oil prices, due to the effect on supply rate, and increasing transport costs
  - leaving tankers vulnerable to terrorist attacks and theft by pirates
  - increasing the risk of shipping accidents and oil spills
- Blocking the Strait of Hormuz for 1 month would reduce the US GDP by about $15 billion per quarter.
- The **Suez Canal** reduces distance and journey times and therefore costs for ships travelling from the east to Europe and America.
- The canal is operated and maintained by the Egyptian-owned Suez Canal Authority.
- In 2017 Egypt collected $5.3 billion in revenue from canal tolls.
- Its importance may diminish. Melting Arctic ice may result in ships travelling north along Russia's northern coast, which is a shorter and faster route. This could increase Russia's strategic importance.
- In 2017 Panama collected $1.6 billion in revenue from the **Panama Canal**.
- An agreement with Panama gives the USA the right to use military force to defend the canal against any threat to its neutrality.
- $5.2 billion has been spent enlarging the canal to take larger ships.

Piracy hotspots

- In recent years there has been an increase in piracy, putting ships at risk.
- While attacks are often on small boats, merchant shipping has also been attacked. Pirates may steal cargoes or hold hostages or boats to ransom.

- Piracy can also increase the risk of an environmental disaster. Crews of merchant ships are often locked up and the ship left unmanned, risking accidents with possible oil spills.
- Hotspots are the Indian Ocean, Strait of Malacca and the coast of Somalia.
- The economic cost of Somali piracy was $7 billion in 2010 but reduced to $1.7 billion by 2016.
- Costs of anti-piracy operations in the Indian Ocean was $1.5 billion in 2016.
- The reduction in piracy has partly been due to:
  - boats protecting themselves with razor wire and armed guards to prevent pirates boarding
  - NATO, EU and countries such as Russia and India using warships to patrol the Indian Ocean

The UK's past as a maritime power

- As an island nation, the UK has always connected with countries around the globe to trade.
- With the growth of the British Empire connections developed and the UK grew into a marine superpower to protect its colonies.
- Many connections continue with countries belonging to the Commonwealth.
- In the past, UK cities with a maritime heritage became culturally diverse due to trade and the arrival of immigrants, for example London, Liverpool, Bristol and Cardiff.

Now test yourself

TESTED

13 How does possession of an island territory help a nation extend its ownership of ocean resources?

14 Why is it important to have secure control of oil chokepoints?

Answers on p. 172

Global flows of shipping and sea cables

REVISED

Changing trends, patterns, networks and regulation of shipping

- 90% of the world's trade is carried by sea.
- The growth in trade resulting from globalisation has led to a 400% increase in the number of ships at sea since the 1990s.
- The biggest increase in shipping was in the Indian Ocean and seas around China.
- World seaborne trade reached 10.6 billion tonnes in 2017.
- Developing countries accounted for 59% of exports and 64% of imports.
- Asia is an important source of exports to Europe and North America (Figure 4.8).

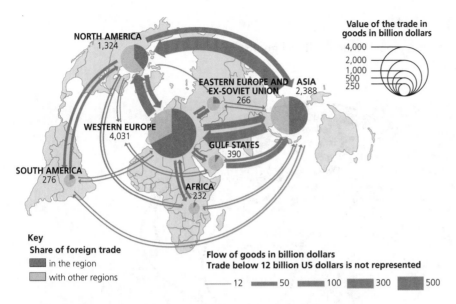

Figure 4.8 Global trade of goods 2015

Table 4.13 Trends in container shipping and oil tankers

Container shipping	Oil tankers
• **Container** ships carried 60% of seaborne trade, worth $12 trillion in 2017. • The quantity of goods carried has risen from 100 million tonnes in 1980 to 1.7 billion in 2016. • Ships have increased in size. In 2000 the largest ship carried 8,000 containers. This had increased to over 18,000 by 2018. • The increased capacity of container shipping is greater than the growth in trade, resulting in the large Hanjan Shipping Company going bankrupt in 2016.	• Oil shipping accounts for 30% of global maritime trade. • Capacity of the tanker fleet has increased 73% since 2000. • The size of a tanker is determined by the canals and chokepoints it passes through (p. 119). • The amount of oil transported can be very dependent on the price of oil. In 2015 oil prices were relatively low and China stockpiled oil, using some tankers for storage. • China creates a significant demand for tankers because it is a major importer of oil.

Containers (intermodal containers) are standard-size steel containers used to transport a huge variety of goods. They can be easily transferred between different methods of transport without being opened, making trade cheaper and more efficient.

Regulation of shipping

● UNCLOS allows ships the right of peaceful passage through territorial waters.
● The International Maritime Organization (IMO), a UN agency based in London, regulates the shipping industry with responsibility for the safety of life at sea and protection of the marine environment. Member governments are responsible for enforcing the regulations.
● The IMO rules require oil tankers to have double hulls to help prevent spillages in the event of an accident.
● To reduce oil pollution tankers are prohibited from flushing out empty tanks with sea water and pumping it into the sea. Sea water used as ballast when travelling empty must be carried in a segregated ballast tank and not in the empty oil tanks.

Smuggling and people trafficking

As well as legal trade, there is a growing global trade in illegal goods and the movement of people (Figure 4.9).

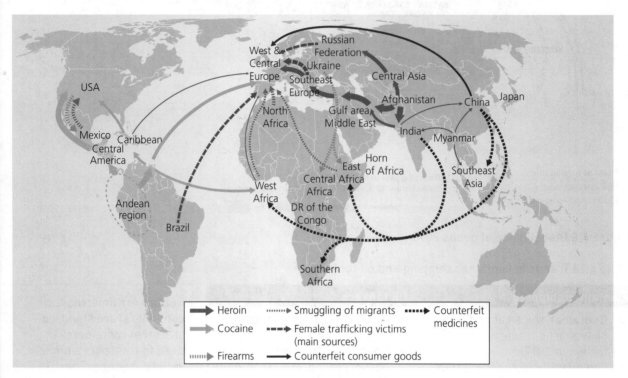

Figure 4.9 Transnational organised crime flows (Source: UN)

- A result of globalisation has been the diversification and international nature of illegal trade.
- Globalised crime includes marketing of illegal and counterfeit goods across continents and the smuggling of migrants and trafficking of people.
- Globalisation of trade has led to deregulation, which smugglers can exploit.
- Globalisation has increased the inequality between places which, combined with conflicts, encourages people to leave. Restrictive migration policies result in people attempting to enter countries illegally. Migrants relying on organised crime are vulnerable and can end up in modern-day slavery.
- Human trafficking generates $30 billion a year worldwide, with an estimated 2.4 million victims from 127 countries.
- There is a lack of international cooperation or solutions to control smuggling and human trafficking. Most attempts to control smuggling are by strengthening border controls.

Growth of seafloor cable data networks

Seafloor cables capable of sending messages have linked continents for over 150 years. Modern cables are cheaper, and carry data faster, than using satellites.

- The increased use of digital technology and the demand for data has increased the demand for global connectivity. Increasing global access to computers is raising demand by 40% a year.
- The distribution of seafloor cables shows inequalities between continents, reflecting population size, access to digital technology and the demand for data transfer. Europe, North America and Asia have the greatest data flow requirements.

- Most seafloor cables are owned by MNCs, often working as a consortium of companies because of the high costs involved. For example, the SEA-ME-WE 5 cable, stretching 20,000 km and linking Southeast Asia, the Middle East and Western Europe, involves 18 operators based in the three continents.
- Google, Microsoft and Amazon are major investors in new cables.

Risks to seafloor cable data networks

There are a number of risks to seafloor cables (Table 4.14). Cables are thicker and more protected (some are buried) in shallow water compared with those on deep ocean floors, where the risk of damage is less.

Table 4.14 **Risks to seafloor cables**

Earthquakes	Can create undersea landslides and rapid, downhill flows of water (turbidity currents), which damage cables. For example, on 26 December 2006 a magnitude 7 earthquake off the coast of Taiwan caused a cable break and currents caused 26 faults, which took 49 days to repair. Data traffic was lost at first but then slowed due to re-routing through other cables.
Tsunamis	Cables can be damaged by seabed erosion and landslides created by the tsunami. For example, the tsunami created by the Tōhoku earthquake in 2011 damaged half of the cables crossing the Pacific Ocean, slowing down data transfer to the USA.
Meteorological events	Storm surges and rainfall from tropical storms can damage cables close to the shore. For example, in 2009 Typhoon Morakot caused an increase in river discharge, which carried mud into an underwater canyon, breaking two cables.
Current abrasion	6% of faults are caused by water movements scraping cables against rocks.
Fishing and anchors	Caused by trawler nets and anchors dragging on the seabed, catching cables. This accounts for 70% of all faults.
Intentional cable cut	Sabotage is rare due to difficulties in accessing cables and the dangers of high voltages. In March 2003 three men were arrested for attempting to cut a cable off the coast of Egypt, which resulted in a 60% drop in internet speeds.

- There are about 100 seafloor cable faults a year. Where water depth is over 1,000 m, most faults are caused by natural events.
- Data can be re-routed through undamaged cables, reducing the impact of damage.

International conventions to protect seafloor data cables

- The first treaty on submarine cables was in 1884.
- UNCLOS states that members must use domestic legislation to penalise cable damage by ships in their jurisdiction.
- Members are free to lay cables in their EEZ and on the continental shelf.
- States can establish no fishing or anchoring zones around cables.
- Cable owners are allowed to use radar tracking systems to monitor ship movements and can then warn ships if they are close to cables. These are sometimes avoided because they could give information to saboteurs.

Now test yourself

TESTED

15 How has the use of containers increased the speed at which goods can be transported around the globe?
16 Why are most seafloor cables owned by MNCs?

Answers on p. 172

Sovereignty of ocean resources

Distribution and ownership of ocean resources

The ocean's resources, which include food, fuels, minerals, sand and gravels and renewable energy, have great economic importance.

- The EEZs defined by UNCLOS established each country's rights to control seabed resources (p. 117).
- UNCLOS designated high-sea resources as part of the 'heritage of mankind' and decided that the benefits should be shared by all countries. It established the International Seabed Authority (ISA), which regulates deep-sea mining.

Mineral resources

- Sand and gravel are the most mined materials in the marine environment. They are extracted from shallow coastal waters in many parts of the world to be used in the construction industry and coastal management (p. 30).
- Minerals washed from the land are extracted from shallow waters, for example:
  - diamonds off the coast of South Africa and Namibia
  - tin and titanium along parts of the African and South American coasts
  - gold off the Alaskan coast
- Manganese nodules, which are rocks that also contain iron, nickel, copper, titanium and cobalt, are considered the most important mineral deposits in the sea. The largest deposit is in the Pacific Ocean over an area the size of Europe between Mexico and Hawaii. They are also found 3,000 km off the coast of Peru and 2,000 km off the east coast of Australia, as well as in the Indian Ocean. While harvesting is possible, the depth of the nodules' location (3,500–6,000 m), the lack of a commercial mining machine and falling mineral prices have meant that mining is not economically viable at the moment.
- Cobalt crusts are metallic layers on the edge of submarine volcanoes, over half being in the Pacific Ocean. The crusts also contain magnesium and iron. The depth of water and the need to separate the crust layer has meant that the minerals are not yet exploited.

Fossil fuels

- Around one-third of all oil and gas extracted comes from offshore sources (Figure 4.10).
- Areas that are economically viable to extract, known as reserves, are found in many parts of the world.
- As many shallow-water reserves, such as the North Sea and coastal areas around the USA, are being exhausted, companies are moving into reserves in deeper waters.
- Technological advances have resulted in the discovery of over 400 major deposits in deep-water locations, such as the Santos Basin off Brazil's coast.

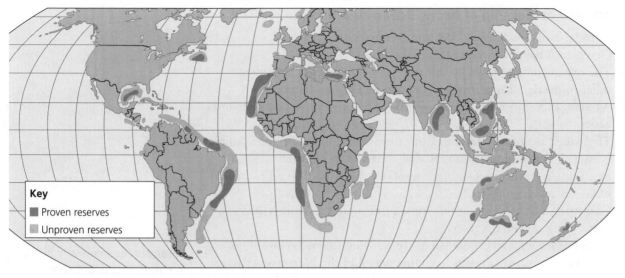

Figure 4.10 Offshore reserves of fossil fuels

Geopolitical tensions and contested ownership

- Conflicts can arise where EEZs overlap and countries disagree over the ownership of islands (p. 118).
- China has claimed island groups (including some man-made islands) and associated EEZs in the South China Sea (Figure 4.11).
- This gives China access to possible resources and control over important oil transportation routes.
- Surrounding countries, especially the Philippines, contest the claims.
- In 2016 the UN declared the man-made islands too small to have EEZs and found in favour of the Philippines. China declared the decision unlawful.
- 78 countries, including Australia, Canada and India, want to extend their rights to ocean resources to the edge of the **continental shelf**, extending their EEZs and creating further contested ownership issues.

> The **continental shelf** is a broad, relatively shallow, gently sloping section of seabed, which is part of the continental crust.

Revision activity

Complete an internet search to find images of the development of the islands in the South China Sea claimed by China. Does this development strengthen its claim to the ocean surrounding the islands?

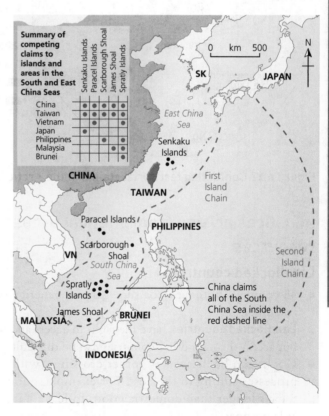

Figure 4.11 Competing claims for islands in the South and East China Seas (Source: Cameron Dunn)

Contested ownership of Arctic Ocean resources

- Canada, Norway, Russia, USA and Denmark (as a result of Greenland) have an Arctic Ocean coastline.
- The ocean contains possibly vast reserves of oil and gas, as well as other minerals, such as iron ore, copper, nickel and diamonds.
- As the sea ice cover diminishes, access to the mineral reserves becomes easier.
- The Arctic nations are laying claims to parts of the ocean, for example Russia claims the Lomonosov Ridge as part of its continental shelf, extending its territory. Denmark wants to prove that it is an extension of Greenland (Figure 4.12).
- Countries such as China are interested in resources, while MNCs are researching oil and gas reserves.
- With less sea ice, travel companies are wanting to send cruise ships into the area. Canada wants to claim control of the Northwest Passage as an important route from the Atlantic Ocean to the Pacific Ocean. The USA disputes Canada's claim.

> A **landlocked country** is a nation that is entirely surrounded by land.

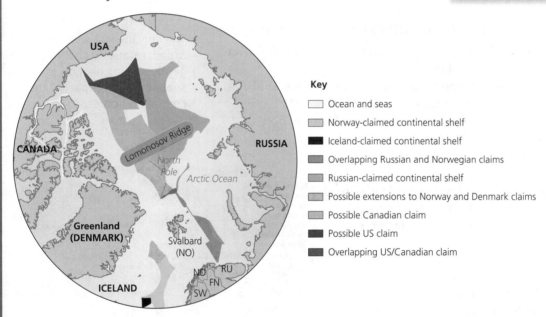

Key
- ☐ Ocean and seas
- ▨ Norway-claimed continental shelf
- ■ Iceland-claimed continental shelf
- ▨ Overlapping Russian and Norwegian claims
- ▨ Russian-claimed continental shelf
- ▨ Possible extensions to Norway and Denmark claims
- ▨ Possible Canadian claim
- ■ Possible US claim
- ▨ Overlapping US/Canadian claim

Figure 4.12 Conflicting territorial claims in the Arctic (Source: Cameron Dunn)

Injustices arising from unequal access to ocean resources

Landlocked countries

- 45 countries without a coastline gain no benefit from ocean resources.
- **Landlocked countries** have the right to access and utilise the oceans, but access through other countries can be difficult.
- Goods can be delayed at border crossings, with clearance processes adding to the cost. For example, goods for Uganda take 5 days longer than domestic goods to leave Tanzania's Dar Es Salaam port.
- Difficulties in trading can lead to slower economic growth — 16 landlocked countries are among the poorest in the world.

> **Typical mistake**
>
> Remember that there will be exceptions. In some countries with a coastline there will be parts of the country that are further inland than some landlocked countries, and thus may not benefit from the ocean. Some landlocked countries may benefit from the presence of large rivers, which allow container-carrying vessels to navigate inland, facilitating trade.

Indigenous people

- Fishing and seafood are an important part of the culture of coastal indigenous people in many places, including North America, Africa and the Arctic.
- Off the African coast, commercial fishing boats have encroached on native fishing areas, with impacts on the local population. For example, in Madagascar community fishing stocks have been almost destroyed.
- Mineral extraction by foreign companies can impact on the activities of local people.
- Indigenous groups in the Brunswick Sea region of Papua New Guinea petitioned the government over fears concerning the impact the Canadian company Nautilus Minerals' deep-sea mining activities would have on the environment and their livelihoods.

Now test yourself

TESTED

17 Why do many countries want to extend their ocean ownership to the edge of the continental shelf?

Answer on p. 172

Managing marine environments

REVISED

The global commons

- The high seas represent one of the four **global commons**.
- The concept of the 'tragedy of the commons' (TOC) is that common resources will be over-exploited unless there is some form of regulation.
- Despite each nation having responsibility for its EEZ, there is a need for international cooperation in sustainable management as fish stocks and pollution move between EEZ boundaries and the high seas.
- The importance of sustainable management is reflected in the UN Sustainable Development Goal 14 — 'Conserve and sustainably use the oceans, seas and marine resources for sustainable development'.
- Concerns over human activity and its impact on the oceans has been responsible for demands to protect the marine environment, for example the eventual ban on whale hunting by the International Whaling Commission and the recent demands for action to control plastic pollution.

> The **global commons** are the global-scale, natural assets that are outside the jurisdiction of any nation. They include the high seas, atmosphere, Antarctica and outer space.
>
> **Over-exploitation** involves harvesting species from the wild at a rate faster than the natural population can recover.

Revision activity

Why do some countries still hunt whales despite the ban by the International Whaling Commission?

Causes and consequences of over-exploitation of marine ecosystems

An example of **over-exploitation** of a marine resource is overfishing, which has occurred due to a number of factors:

- Increasing populations leading to increased demand for food.
- A global doubling in fish consumption, from about 10 kg per person in 1960 to over 20 kg in 2016. Much of this increase results from increased wealth and changing tastes in China.
- Increase in fishing fleets. Over 55% of the oceans are fished industrially by over 70,000 fishing vessels. Factory fishing boats are able to stay at sea longer, processing and freezing catches.
- Improvements in technology allow fishing boats to find and track shoals of fish that previously would have gone unnoticed.

- In the high seas there are limited rules regarding fishing practices. In regulated areas, the size of the fishing grounds and the lack of resources make enforcing regulations difficult.
- Some large-scale fisheries receive government subsidies, which result in more boats than are needed to meet demand.
- Only 1.5% of the oceans are protected areas and most of these are still open to fishing, which can deplete fish stocks.

Consequences of overfishing

As well as affecting the marine ecosystem, overfishing can have an impact on peoples' lives:

- Fish provides the world's population with 15% of its dietary animal protein. In many developing countries fish is an important, affordable food. As fish stocks decrease, prices may rise out of the reach of poorer members of the population.
- As fish prices rise, more developing countries export their fish for international trade, leaving only locally caught fish available to the population.
- In developing countries in Africa and Asia, local fishermen provide the bulk of fish for poor communities. Competition from commercial fishing fleets has resulted in small fishing operations in Southeast Asia being taken over by large companies, leading to loss of livelihood, especially in rural areas where there is little alternative employment.
- Decline in fish stocks can result in unemployment in fishing and associated industries such as processing.
- In Senegal, West Africa, there is 80% unemployment among Senegalese fishermen due to a decline in fish stocks.
- The collapse of the Newfoundland cod fishery in Canada in 1992 made 40,000 people unemployed.
- In the EU, the Common Fisheries Policy sets quotas for how much of each type of fish the member countries can catch, in an attempt to manage fish stocks. This has been blamed for unemployment and affecting the livelihood of fishermen in places like Scotland.
- Consumers have had to accept alternative species of fish, for example hake or pollack instead of cod.

Sustainable management of marine environments

- The aim of sustainable management is to safeguard marine resources for future generations while allowing for global growth and stability. It usually involves controlling economic activity in an attempt to protect wildlife or allow fish stocks to recover.
- Despite UNCLOS and the UN Sustainable Development Goals, there is no enforceable global policy on sustainable management of the oceans.
- A number of strategies has been implemented by different organisations (Table 4.15).

Exam practice answers and quick quizzes at www.hoddereducation.co.uk/myrevisionnotesdownloads

Table 4.15 Methods of sustainable management of marine environments

No-catch zones	Form part of no-take zones, which are marine areas where the removal of any resources is prohibited There are three zones in the UK — Lundy, Flamborough Head and Lamlash Bay — where taking fish is prohibited for reasons of nature conservation
Marine Protected Areas (MPAs)	Areas of the marine environment that have been reserved by law to protect part or all of the natural resources within them They can limit fishing, mining and tourism Most are found within territorial waters where the government involved can enforce them The Great Barrier Reef off the Queensland coast of Australia is an MPA
Marine Conservation Zones (MCZs)	The 2009 UK Marine and Coastal Access Act allowed the creation of MCZs They are designed to protect a range of nationally important marine wildlife and habitats There are 50 MCZs in the seas around England protecting over 20,000 km^2 of sea, including 'Offshore Brighton' in the English Channel, which covers a diverse range of species living in an area of 862 km^2
Fishing quotas	To reduce catches to a biologically and economically sustainable level authorities will introduce Total Allowable Catches (TACs) Quotas are based on scientific advice on fish stocks The EU, USA and New Zealand employ quotas

Managing ocean pollution

REVISED

Sources and causes of ocean pollution

- Marine pollution can be both deliberate and accidental.
- Up to 80% of marine pollution originates from land (Table 4.16).

Table 4.16 Sources and causes of marine pollution

Source	Examples
Terrestrial runoff	Water running from the land can pollute the oceans by carrying many contaminants: • fertilisers, pesticides and particles rich in nitrogen, and phosphorous from agricultural land • oil from road surfaces • waste from industrial activity and inland mining, which may contain toxic chemicals • treated and untreated sewage • domestic waste water, which may contain chemicals from products used in the home and plastic microbeads used in toiletries • litter, especially plastic products, which get blown or washed into rivers and carried to the sea
Waste disposal	As well as sewage disposal, other materials are disposed at sea: • Rubbish is dumped from shipping. In the North Sea up to 40% of marine litter comes from shipping, creating 20,000 tonnes of waste. • Over 80% of all waste dumped at sea is material that has been dredged from elsewhere.
Oil spills	Oil spills can result from shipping accidents and also from leaks from extraction activities: • Tanker accidents have reduced from an average 24.5 per year in the 1970s to 1.7 since 2010. • Between 2000 and 2017 there were 53 spills from tankers, resulting in 47,000 tonnes of oil being lost. • Leaks can occur from seabed oil wells. In 2010 an explosion in the BP oil rig Deepwater Horizon in the Gulf of Mexico led to a 3-month oil spill that allowed over 6,000 tonnes of oil a day to escape.
Pollution from the atmosphere	The ocean naturally absorbs carbon dioxide. As the amount of carbon dioxide in the atmosphere increases due to climate change, the oceans absorb more and become more acidic (p. 94).

Consequences of ocean pollution

Eutrophic dead zones

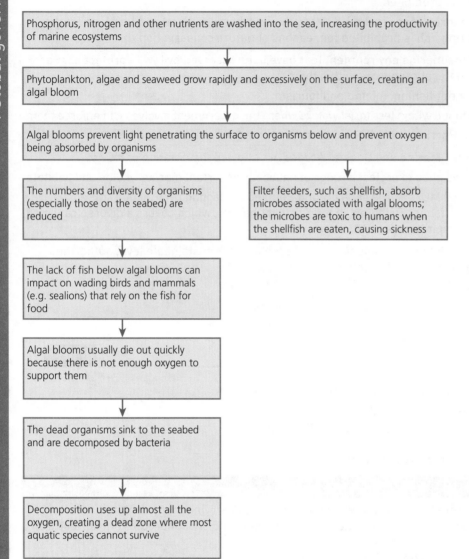

Phosphorus, nitrogen and other nutrients are washed into the sea, increasing the productivity of marine ecosystems

↓

Phytoplankton, algae and seaweed grow rapidly and excessively on the surface, creating an algal bloom

↓

Algal blooms prevent light penetrating the surface to organisms below and prevent oxygen being absorbed by organisms

↓ ↓

The numbers and diversity of organisms (especially those on the seabed) are reduced

Filter feeders, such as shellfish, absorb microbes associated with algal blooms; the microbes are toxic to humans when the shellfish are eaten, causing sickness

↓

The lack of fish below algal blooms can impact on wading birds and mammals (e.g. sealions) that rely on the fish for food

↓

Algal blooms usually die out quickly because there is not enough oxygen to support them

↓

The dead organisms sink to the seabed and are decomposed by bacteria

↓

Decomposition uses up almost all the oxygen, creating a dead zone where most aquatic species cannot survive

Figure 4.13 The formation of eutrophic dead zones

The Baltic Sea has seven of the 10 largest marine **dead zones** due to increased runoff of agricultural fertilisers and sewage, as well as over-fishing of cod. The cod were at the top of a food chain that started with algae being eaten.

Plastic garbage patches

- The massive increase in the use of plastics globally has led to large amounts finding their way into the oceans.
- Plastic bottles can take 450 years and fishing line 600 years to break down.
- UV rays and the ocean environment break the plastic down into pieces smaller than $1\,cm^2$.
- Scientists have discovered that degrading plastics release toxic chemicals such as bisphenol into the ocean.

> **Dead zones** are areas of ocean with such low concentrations of oxygen that they are unable to support marine life. While they can occur naturally, they are often caused or enhanced by pollution from human activities.

- Circulating ocean currents called **gyres** (Figure 4.14) consolidate debris into certain parts of the world's oceans known as garbage patches.

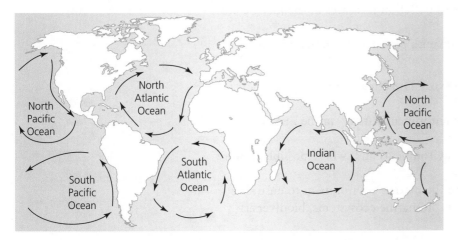

Figure 4.14 Global pattern of gyres

The North Pacific has created the Pacific garbage patch, which is:
- over 1.6 million km² (three times the size of France)
- estimated to contain 1.8 trillion pieces of plastic (250 pieces for every person on the planet)
- largely made up of discarded and lost fishing gear (over half)

Plastic waste endangers sea life:
- Animals can get entangled, restricting movement and causing starvation or injuries that become infected. Dolphins and turtles get caught up in lost fishing nets.
- Seabirds ingest plastic, often mistaking it for food. Plastic is found in the stomach of 80% of seabirds, with the highest concentrations around south Australia, South Africa and South America.

Oil spills

Seabirds land on oil spills and get covered in oil:
- When oil gets on to the feathers it prevents birds from flying.
- It destroys the waterproofing and insulation, resulting in death from hypothermia.
- Birds preening their feathers when trying to recover ingest oil, which can damage internal organs, leading to death.

Other animals are less vulnerable, but can be affected:
- The blowholes of whales and dolphins become clogged, affecting breathing.
- Oil coats the fur of otters and seals, making them vulnerable to hypothermia.
- Fish can be poisoned by oil, especially once it has become a thin emulsion in the water.

The Exxon Valdez oil spill in 1989 killed up to 500,000 seabirds, thousands of otters, hundreds of seals and 24 killer whales. Billions of salmon and herring eggs were destroyed, and fish stocks have still not recovered.

Revision activity

Research what action is taken to deal with a major oil spill.

Strategies to manage marine waste

UNCLOS makes it an obligation of all states to protect and preserve the marine environment:

● Each coastal state can enforce its pollution measures in the marine area of its EEZ.
● States are required to cooperate with global and regional efforts to combat pollution.

The EU's Marine Strategy Framework Directive requires all members to take necessary measures to:

● achieve and maintain 'good environmental status' of the marine environment by 2020
● protect and preserve marine environments and, where possible, restore marine ecosystems
● prevent and reduce inputs into the marine environment to ensure there are no significant impacts on marine ecosystems, biodiversity and human health

The UK has some policies that aim to reduce plastic waste entering the oceans:

● Microbeads have been banned in toiletries.
● The charge for carrier bags intended to reduce waste going to landfill has reduced bag usage and so decreased the chance of bags entering the ocean.
● In 2018 the government promised £61.4 million to deal with plastic pollution in the oceans. This included:
  ○ £25 million for research
  ○ £20 million used to reduce plastic and other pollution entering the seas from manufacturing in developing countries
  ○ £16.4 million for improving waste management to stop plastics entering water
● Many NGOs, such as Friends of the Earth and Greenpeace, have attempted to raise the issue of plastic pollution.
● In 2017 the final episode of the BBC TV series *Blue Planet 2* raised awareness in the UK of the issue, leading to demands from many individuals for something to be done.

The Ocean Cleanup Foundation, a non-profit foundation set up in 2013 by 18-year-old Boyan Slat, aims to develop technologies to remove plastic from oceans. It aims to clear half of the Pacific garbage patch within 5 years.

Local actions include individuals organising beach clean-ups to reduce plastic waste.

> **Exam tip**
>
> Make sure that you are up to date on the progress being made by the Ocean Cleanup Foundation.

> **Exam tip**
>
> Plastic waste has become a major cause for concern as people have become more aware of the issue. Make sure that you are up to date with the latest attempts to deal with the problem.

> **Exam tip**
>
> Make sure you know one case study of an ocean issue that illustrates the different geographical scales of governance, such as regional, national and international, and how they interact with each other. The example below is of the governance of a UNESCO Marine Heritage Site. You may have studied another example, such as global strategies for conservation of the Arctic Ocean.

Case study: Governance of a UNESCO Marine Heritage Site

The Belize Barrier Reef Reserve System

- UNESCO has recognised 49 Marine Heritage Sites in 37 countries.
- In 1996 it made the Belize Barrier Reef Reserve System a Marine Heritage Site.

- The area off the east coast of central America is a 300 km stretch of coral reefs (p. 22) ranging between 300 metres and 40 kilometres off the coast.
- The area is Belize's top tourist attraction, with over 130,000 visitors a year, and is also vital to the country's fishing industry.
- A number of stakeholders have had roles in the protection of the site (Table 4.17).

Table 4.17 Roles of stakeholders in the governance of the Belize Barrier Reef Reserve System

Stakeholder	Role
UNESCO	Added the Belize Barrier Reef Reserve System to the Marine Heritage Site list
	In 2009 it added the area to its 'in danger' list, stating that better management and safeguards were needed
Belize government	Protects the area under its National Protected Area Policy
	In 2009 laws were introduced to protect endangered species of fish, such as parrot fish, and spear fishing was banned
	Banned offshore drilling for oil in the area in 2017
NGOs	The Wildlife Conservation Society provided research to help inform the changes in fishing policy
	WWF helped put public pressure on Belize to ban oil extraction
Local communities	Government training allows locals to be tour and diving guides — local guides must be used by law
	Involvement persuades local communities to protect the reef as an asset
Fisheries	Supported a licensing system that controls how much, and where, fishing is allowed
	Fisheries must provide catch data to monitor production and assist management decisions
	Play an active role in setting out policy and carrying out enforcement

Now test yourself

TESTED ☐

18 How are fishing quotas an example of sustainable management?
19 What has been done to reduce the occurrence of oil spills?

Answers on p. 172

Exam practice

The format for the different examination papers is shown below.

Specification	Method of examination
Eduqas A-level Component 2; Section B	Two compulsory data-response questions and one extended-response question
WJEC A2 Unit 3; Section B	Two compulsory data-response questions and one extended-response question

Eduqas A-level and WJEC A2 format

1 Figure 1 shows flows of migrant remittances in 2011

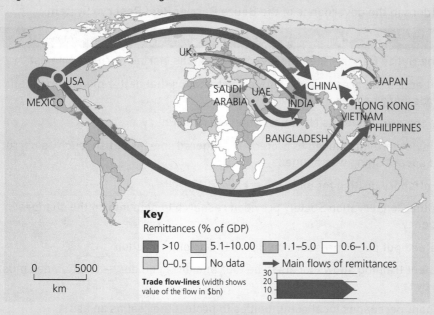

Figure 1 Flows of migrant remittances 2011

a) Analyse the pattern of remittances shown in Figure 1. [5]

b) Explain one benefit and one risk of interdependency for different countries. [5]

2 Figure 2 shows the global distribution of undersea data cables in 2012.

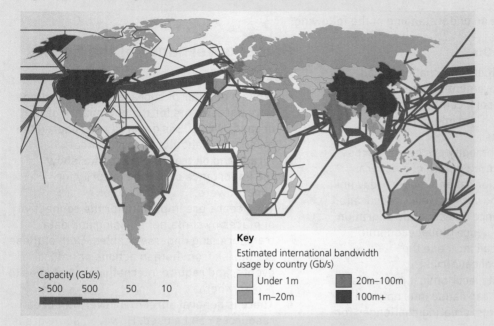

Figure 2 Global distribution of undersea data cables 2012

a) Describe the distribution of undersea data cables shown in Figure 2.   [5]
b) Explain two ways in which superpower nations have contributed to the global
    governance of oceans.   [5]

Eduqas A-level format

3 'Globalisation has led to an increasingly borderless world.' Discuss.   [20]

WJEC A2 format

4 Evaluate the effectiveness of agreements regulating the use of ocean resources.   [18]

Answers and quick quiz 4 online

ONLINE

Summary

You should now have an understanding of the following:

Global migration

- Globalisation includes international trade, migration and movement of data.
- Changes in transport, communications and media have led to a 'shrinking world' for potential migrants.
- Poverty, poor commodity prices and access to markets, the development of diaspora communities, historical and modern-day links, as well as superpowers, have all encouraged and had an influence on economic migration.
- Migration can influence global economic inequalities and can increase the interdependency of countries.
- Geopolitical events, economic injustices and natural disasters can cause international refugee movements, creating challenges for host nations.
- Push factors in rural areas and employment pull factors in urban areas have encouraged rural–urban migration in many developing countries.

Global governance of the Earth's oceans

- While a number of supranational institutions have agreements for the governance of oceans, UNCLOS is generally considered the most important global governance.
- Piracy and oil transit chokepoints show the importance of marine security for superpowers.
- The oceans are important for the connectivity of places by container shipping and data transfer using undersea cables. Both of these are at risk from human actions or natural events, and require international strategies to protect them.
- There is general agreement over how ocean resources can be used. However, some areas are claimed by more than one nation, creating tensions.
- Landlocked countries have unequal access to ocean resources.
- Over-exploitation of marine ecosystems requires sustainable management to help provide global growth and stability.
- Marine pollution is an increasing problem with many causes. It requires enforceable strategies and international cooperation to deal with the issue.

5 Tectonic hazards

Tectonic hazards

Tectonic processes and hazards

Characteristics of the Earth's structure

- The Earth can be divided into three layers, based on their density and chemical composition (Figure 5.1 and Table 5.1).

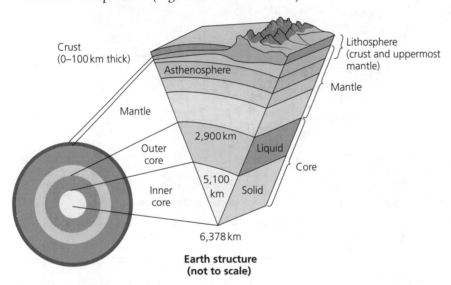

Figure 5.1 The structure of the Earth

Table 5.1 Composition of the Earth's layers

Layer	Characteristics
Core	• Consists of iron and nickel at temperatures of 6,200°C • Inner core is solid, 1,250 km thick • Outer core is semi-molten, 2,200 km thick
Mantle	• Consists of silicate rocks, rich in magnesium, making up 84% of the Earth's volume • Up to 2,900 km thick, with temperatures of 1,000°C near the crust and 4,000°C near the core • Rocks are semi-molten • Asthenosphere: – The upper part of the mantle at depths of 80–200 km – Although solid, the rocks can deform under pressure
Crust	• Thin outer layer of the Earth, divided into the following: – Oceanic crust, 6–10 km thick, made up of basaltic rocks composed mainly of silica and magnesium, known as sima – Continental crust, up to 70 km thick, made up of granitic rocks composed mainly of silica and alumina, known as sial

- The **lithosphere** is the rigid outer layer of the Earth consisting of the crust and solid outermost layer of the upper mantle.
- The **Moho discontinuity (Moho)** is the boundary between the crust and the mantle.

The mechanisms of plate movements

The lithosphere is divided into seven large and three smaller plates, with a number of minor plates (Figure 5.2). Plates can move:

- towards each other at **converging/destructive plate margins**, for example off the west coast of South America
- apart from each other at **diverging/constructive plate margins**, for example Iceland
- alongside each other in opposite directions, or the same direction at different speeds, at **conservative plate margins**, for example the San Andreas Fault in California

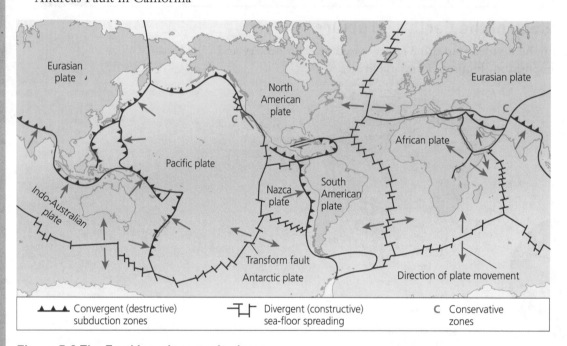

Figure 5.2 The Earth's main tectonic plates

There are three competing theories to explain the main drivers of tectonic plate movements.

Internal heating and mantle convection currents

- Heat from the core of the Earth is moved towards the surface by convection currents in the mantle.
- These spread out at the surface, carrying the plates with them like a conveyor belt.
- The plates move on the asthenosphere — a layer between the upper mantle and the lithosphere.
- Modern scientific techniques have been unable to identify any convection cells in the mantle that would be powerful enough to move plates.

Ridge push

- Molten magma rises at a mid-ocean ridge divergent boundary.
- Rocks in the lithosphere are heated, expand and rise above the surrounding seafloor, forming a slope away from the ridge.
- As the rock cools it becomes denser and gravity causes it to slide down the ridge, exerting a force on the plates.
- This gravitational sliding is the active driving force.

> **Typical mistake**
>
> The Earth's tectonic plates are made up of the lithosphere and not just the Earth's crust.

Slab pull

- At many convergent plate boundaries one plate is denser and heavier than the other.
- The denser plate subducts beneath the less dense plate.
- The subducting plate is colder and heavier than the mantle, so it continues to sink, pulling the rest of the plate with it.

Recent research suggests that slab pull is the major driving force behind tectonic plate movements.

Processes operating at plate margins

Diverging plate margins

- Diverging plates under oceans, driven by slab pull, bring magma to the surface to form a ridge (Figure 5.3).
- The mid-Atlantic ridge separating the South American and African plates is an example, rising up to 3 km above the ocean floor. The plates are spreading at an average of 2.5 cm a year.
- Volcanoes form along the ridge and may rise above sea level, creating islands. For example, Surtsey, 32 km from Iceland, formed between 1963 and 1967 from volcanic eruptions at the mid-Atlantic ridge.

5 Tectonic hazards

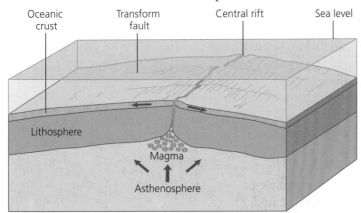

Figure 5.3 Cross-section of a diverging plate margin beneath the ocean

- Where continental plates diverge, mountain ridges are found at the plate margins as hot magma forces the crust to bulge.
- As plates move apart the brittle crust fractures, forming parallel faults.
- Areas of crust slowly subside between the faults to form a **rift valley** between the rift mountains (Figure 5.4).
- The East African Rift Valley is an example, stretching 3,000 km as a result of the African plate splitting in two and moving apart at 7 cm a year.
- Volcanic activity can occur along the plate margin, for example Mt Kilimanjaro in Tanzania, which lies on the eastern edge of the East African Rift Valley.

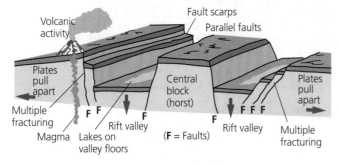

Figure 5.4 The features of a rift valley

Converging plate margins

- Where two oceanic plates converge, the colder, denser plate will subduct beneath the other, forming an ocean trench (Figure 5.5).
- The subducted plate is heated and melts around 100 km below the surface.
- Some molten rock may rise through lines of weakness to the surface, resulting in a curved chain of volcanic islands called an **island arc**.
- The Tonga Islands by the Tonga trench, where the Pacific plate is subducting beneath the Australian plate, is an example of an island arc.

> **Converging plate margins** are also known as destructive plate boundaries.
>
> The **subduction zone** is the place at a converging plate margin where one plate descends below the other.

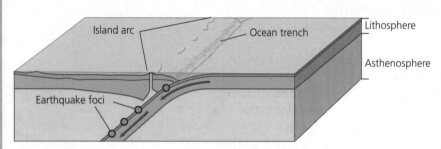

Figure 5.5 Cross-section of an oceanic plate converging plate margin

- Where oceanic and continental crusts converge, the denser oceanic crust is subducted by slab pull (Figure 5.6).
- The less dense continental crust gets buckled and folded, to form mountain chains.
- The oceanic plate melts, forming a magma chamber.
- The magma rises up through weaknesses in the continental crust to form volcanoes.
- The interaction of the two plates in the subduction zone creates earthquake activity up to 700 m deep in an area called the **Benioff zone**.
- The Cascade Range in western North America is an example of mountains formed by the oceanic Pacific plate subducting under the continental North American plate.

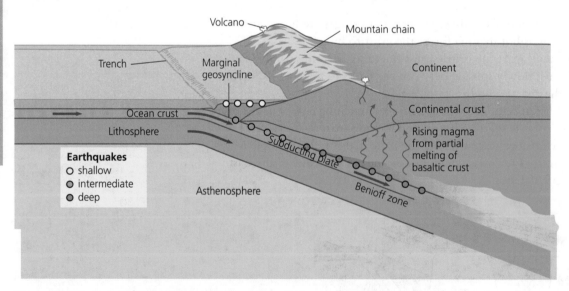

Figure 5.6 Cross-section of an oceanic and continental plate converging plate margin

Exam practice answers and quick quizzes at **www.hoddereducation.co.uk/myrevisionnotesdownloads**

- When two continental plates converge, their similar densities result in no subduction.
- The plate edges and sediments trapped between the plates are forced up, creating fold mountains.
- The lack of subduction means that crust is not melted, so there is no volcanic activity. Earthquakes can occur.
- The Himalayas, formed by the Indo–Australian plate moving 5 cm a year into the Eurasian plate, are examples of fold mountains. This is a long-term process, taking millennia.

Conservative plate margins

- When two plates slide past each other there is no subduction and so no volcanic activity (Figure 5.7).
- Earthquakes occur, and these can be powerful if a build-up of pressure is suddenly released.
- The San Andreas Fault in California, where the Pacific plate is moving 5 cm a year faster in the same direction as the North American plate, is an example.

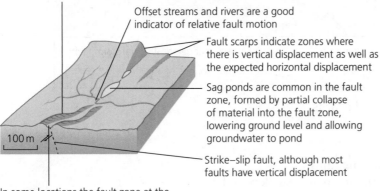

Compression ridges form where the fault has a turn or kink and the ground heaves upwards

Offset streams and rivers are a good indicator of relative fault motion

Fault scarps indicate zones where there is vertical displacement as well as the expected horizontal displacement

Sag ponds are common in the fault zone, formed by partial collapse of material into the fault zone, lowering ground level and allowing groundwater to pond

100 m

Strike–slip fault, although most faults have vertical displacement

In some locations the fault zone at the surface forms a depression.
The area is more easily eroded as the material is broken-up fault gouge

Figure 5.7 The features of a conservative plate margin

Hotspots

- At a **hotspot**, a plume of magma rises from the asthenosphere. The magma pushes through the crust, forming volcanoes at the surface.
- As the plate moves over the hotspot it creates a line of volcanoes, which move with the plate.
- As the oceanic volcanoes move away from the hotspot they cool and subside, remaining as islands or, if under water, sea mounts.
- As continental volcanoes move, they become extinct.
- An alternative theory is that plates have weaknesses from former collisions or are stretched by slab pull. When the weak parts pass over previously subducted material the molten rock is able to move to the surface.
- The Hawaiian Islands, extending 2,400 km, are an example of an oceanic hotspot.
- The islands furthest away from the hotspot formed about 6 million years ago, while those closest to the hotspot are still active volcanoes.
- Yellowstone is an example of a continental hotspot.

The global distribution of tectonic hazards

Figure 5.8 shows how tectonic hazards are generally concentrated along plate margins.

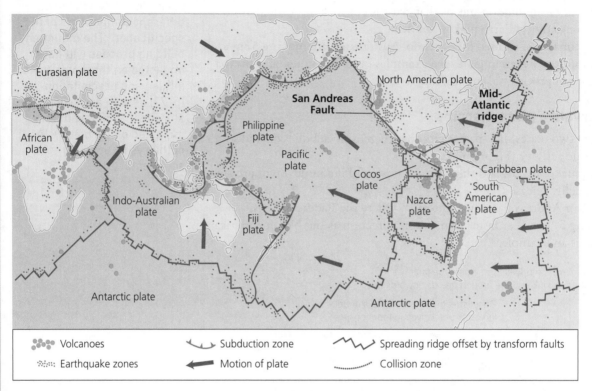

Figure 5.8 The global distribution of tectonic hazards

Earthquakes

The main earthquake zones occur along plate margins:

- Earthquakes along diverging plate margins are mainly shallow and often related to volcanic activity.
- Oceanic crust subduction zones experience frequent earthquakes, which may be powerful and create tsunamis.
- Earthquakes along continental crust converging plate margins are mainly shallow, but can be powerful and hazardous if humans inhabit the area.
- Some earthquakes, including powerful ones, occur away from plate margins often due to stress creating movement along old fault lines.
- Human activity, such as fracking for gas extraction, is thought to cause minor earthquakes.

Volcanoes

Volcanoes are found primarily along plate margins, especially around the edges of the Pacific Ocean where they form the 'Pacific Ring of Fire':

- Divergent plate margins: 75% of magma reaching the surface does so at these plate margins, primarily along mid-ocean ridges. Eruptions are usually non-violent and as most are in the deep ocean they present little hazard to people. The exception is where ocean ridges cross inhabited land, such as Iceland.
- Convergent plate margins: 80% of active volcanoes are at subduction zones. Eruptions can be violent, proving hazardous if close to areas of habitation, for example the Mt St Helens eruption in the Cascade Range in 1980.

Hotspots

Eruptions are usually less violent but produce large quantities of lava. These may be less hazardous but cause damage that can affect people's lives. The eruption on Hawaii's Big Island in 2018 is an example.

The characteristics of the physical hazard profile that influence its impact

The impact of a tectonic hazard depends on a number of physical and human factors.

Revision activity

Make sure you know whether volcanoes, earthquakes or both occur at the different plate margins. Have a named example of a tectonic event that has occurred at each of the different plate margins.

Magnitude

- The magnitude describes the size or physical force of a hazard event.
- Magnitude is considered the most important factor influencing the impact of a tectonic hazard.
- The magnitude of an earthquake can be measured using three different scales (Table 5.2).

Table 5.2 Earthquake magnitude scales

Scale	Description
Richter scale	Based on the amplitude of seismic waves made on a seismograph
	Uses a logarithmic scale from 1 to 9, although there is no upper limit
	Each increase in number represents a tenfold increase in magnitude
	Still widely used
Moment magnitude (MM) scale	Measures the energy released, taking into account factors such as geology, area of the fault surface and amount of movement at the fault
	Uses a scale from 1 to 9
	Used since 2002 by the United States Geological Survey (USGS) for all modern, large earthquakes
Mercalli scale	A qualitative scale that measures intensity based on the effects on the Earth's surface
	Uses a scale from I to XII

- Magnitude of volcanoes can be measured using the **volcanic explosivity index (VEI)**.
- The volume of material erupted, the eruption cloud height, direction of the eruption and observations such as 'gentle' or 'mega-colossal' are used to determine the explosivity value.
- The index ranges from 0 to 8, although there is no upper limit.
- Each increase in number represents a tenfold increase in explosivity. This is a very important point to remember when carrying out quantitative analysis of earthquake data.
- No account is taken of the different densities of erupted materials.
- Sulfur dioxide emissions are not taken into account, so the VEI is limited in describing the effects on the atmosphere.

Exam tip

Although you may see the Richter scale referred to by the media, you should use the MM scale because it is now widely accepted as a better way to measure magnitude.

Predictability

- The unpredictable nature of many tectonic hazards influences their impact.
- The relationship between tectonic hazards and plate boundaries allows the prediction of location but not when an event will occur.

- An accurate prediction can reduce the impact of a hazard by allowing the evacuation of danger zones.
- Too many inaccurate predictions may increase the impact of future hazards as people will begin to ignore warnings.

Earthquakes:
- Scientists have not yet found a characteristic pattern of seismic activity or any other physical, chemical or biological change that indicates a high probability of an earthquake.
- Factors such as increased levels of radon gas, changes in the water table, variations in magnetic fields and the behaviour of animals have all been researched, but none has been found to be a reliable indicator.
- **Seismic gap theory** suggests that, over a period of time, all parts of a fault must attain the same average level of movement. This may be through many minor quakes or the result of a rarer but larger quake. A seismic gap in which no earthquakes occur could be interpreted as indicating increased likelihood of a significant earthquake.

Volcanoes:
- There is greater opportunity to predict a volcanic eruption because there are usually warning signs before the main eruption.
- Scientists can monitor changes in gas emissions, ground deformation, hydrology, temperature changes and seismic activity to help with prediction.
- Despite the warning signs, it is still not possible to accurately predict when an eruption will occur or how large it will be.

Frequency

- In areas where earthquakes or volcanic eruptions occur frequently, the magnitude is usually smaller, thus resulting in less impact.
- Areas with frequent tectonic events may have measures in place to reduce the impact, such as earthquake-resistant buildings and earthquake drills, for example Japan.

Duration

- It can be expected that the longer an event lasts, the greater the impact is likely to be.
- Often aftershocks following earthquakes (e.g. Christchurch earthquake in New Zealand in 2010) and secondary hazards associated with volcanoes, such as lahars (e.g. Mt Pinatubo in the Philippines in 1991) and tsunamis (e.g. Tōhoku in Japan in 2011), extend the duration and increase the impact.

Speed of onset

- Speed of onset refers to how quickly the peak of the hazard event arrives.
- The speed of onset of an earthquake is almost instant, allowing no time to issue warnings and therefore having the potential for a greater impact.
- The build-up to a volcanic eruption and the slower onset can allow evasive action, such as evacuations, reducing the impact. As many as 20,000 lives may have been saved by evacuating the area around Mt Pinatubo in 1991 due to the slow onset of the main eruption.

Areal extent

- The greater the area covered by a tectonic event, the greater its potential impact.
- The impact of an earthquake may be felt for a few kilometres, whereas the impact of gas and ash from a volcanic eruption could be global in extent.

> **Areal extent** refers to the size of the geographical area affected by the tectonic event.

Using the characteristics that influence the impact of a tectonic hazard, it is possible to create a tectonic hazard profile, which allows comparisons between different hazards and different events to be made (Figure 5.9).

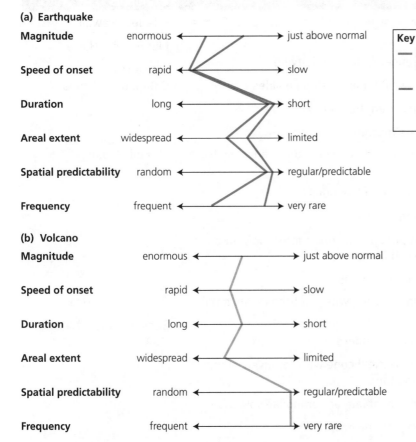

Figure 5.9 Tectonic hazard profiles for an earthquake and a volcanic eruption

The impact of a tectonic event is not only influenced by features of the hazard profile. Population density, level of development and vulnerability of the population are all important factors influencing the impact.

> **Revision activity**
>
> Create hazard profiles for tectonic events you have studied.

Now test yourself

TESTED ☐

1 Using the information in Figure 5.8, why are more people at risk from earthquakes than from volcanoes?
2 Why might a hazard profile similar to those in Figure 5.9 not show the impact a tectonic event may have on an area?

Answers on p. 172

Volcanoes, processes, hazards and their impacts

Types of volcano

Volcanoes can be classified according to their shape and the nature of the eruption (Table 5.3).

Table 5.3 Classification of volcanoes by shape

Type	Shape and formation	Examples
Shield	Effusive eruption where large quantities of lava pour from a central vent Lava travels long distances before solidifying Produces tall volcanoes with gently sloping sides Eruptions tend to be frequent but not explosive Occur at diverging plate margins and hotspots	Kilauea, Hawaii Mauna Loa, Hawaii Fernandina Island, Galapagos Islands
Composite (stratovolcano)	Made of many layers of solidified lava and volcanic ash from different eruptions Lava is slow flowing and solidifies quickly Produces steeply sloping sides Ash occurs in explosive eruptions, often after the vent has become blocked with solidified lava Occur at converging plate margins	Mt Fuji, Japan
Cinder (ash)	Gas forces lava high into the air, where it breaks into small fragments Fragments cool and fall as cinders Produces a short, symmetrical cone with relatively steep sides and a crater at the top Often occur on the sides of shield or composite volcanoes Many are formed from just one eruption	Paricutin, Mexico Sunset Crater, Arizona USA
Acid (dome)	Viscous, acid lava erupts effusively but cannot flow far before solidifying Lava builds up to form a steep-sided cone	Popocatepetl, Mexico Lassen Peak, California USA
Caldera	Gases trapped in a magma chamber eventually cause an extremely explosive eruption The upper part of the volcano can be destroyed The ground above the magma chamber can subside, leaving a large depression, which can eventually flood	Crater Lake, Oregon USA Yellowstone, USA

Eruptions can be classified based on the explosiveness of the eruption, as a result of the crystal and gas content, and the temperature of the magma (Table 5.4).

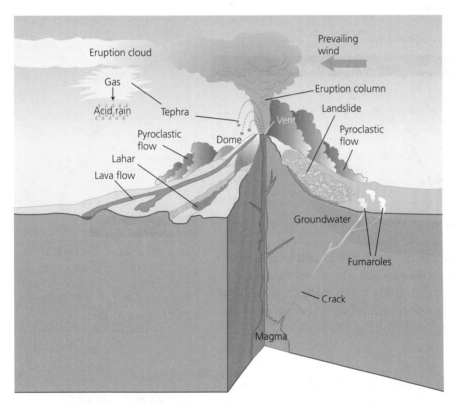

Table 5.4 Classification of volcanoes by type of eruption

Type of eruption			Description
Increasing explosivity	Effusive	Icelandic	Lava flows freely from an open fissure
		Hawaiian	Fluid lava flows from a vent; some may be thrown into the air from the vent
	Eruptive	Strombolian	Distinct eruptions, forcing lava into the air, occur frequently; the least violent explosive eruption
		Vulcanian	Short, relatively small, violent eruption of viscous lava
		Vesuvian	Powerful blasts of gas pushing ash into the air, with some lava flows
		Plinian	Huge, violent eruption blasting ash and fragments high into the air, creating huge ash clouds; gas clouds and lava can rush down the slopes; part of the volcano may be blasted away

Volcanic processes and the production of associated hazards

There are many hazards associated with volcanic eruptions (Figure 5.10).

Figure 5.10 Types of volcanic hazard

Pyroclastic flow

A pyroclastic flow is a high-density mix of hot volcanic gas, volcanic fragments, pumice and ash, which moves at high speed down volcanic slopes. It can form in three ways:

● The **eruption column** collapses. During an explosive eruption, material is ejected into the atmosphere. The gas and particles are denser than the surrounding air and fall back to Earth, flowing down the sides of the volcano.

The **eruption column** is a cloud of hot volcanic ash suspended in volcanic gas from an explosive eruption.

- 'Boiling over' at the vent. Material is emitted from an explosive eruption without being thrown high into the atmosphere and travels rapidly downslope.
- The collapse of a lava dome. The slopes of a volcano become so steep they collapse due to gravity.

Pyroclastic flows are the greatest volcanic hazard:
- The temperature of a flow is usually 200–700°C but can reach over 1,000°C.
- They travel around 80 km/h but can reach over 700 km/h.
- People caught in a flow die immediately from the blast of hot air in front of the flow, which causes extreme burning and asphyxiation.
- A flow can travel many kilometres, destroying everything in its path.
- There is little warning and no time to escape.
- The 1982 eruption of El Chichón in Mexico killed up to 2,300 people and made 20,000 homeless, mainly as a result of pyroclastic flows, which covered 104 km².

Lava flows

Lava flows are streams of molten rock pouring from an erupting vent.
- The speed at which lava travels depends on lava type (Table 5.5) and viscosity, and steepness of the ground.

Table 5.5 Comparison of aa and pahoehoe lava

Feature	Aa lava	Pahoehoe lava
Flow volume	High	Low
Velocity	Up to 10 km/h on steep slopes 1 km/h on gentle slopes	1–2 km/h on steep slopes
Thickness of flow	2–10 m	0.2–2 m
Surface	Rough	Smooth

- Lava flows are the least hazardous process in volcanic eruptions because people have time to move out the way.
- They can impact human life by destroying buildings and infrastructure, and covering agricultural land with rock so that it cannot be used.
- The fastest recorded lava flow was in January 1977 when a lava lake on Mt Nyiragongo in Democratic Republic of Congo drained in less than an hour. Lava flowed at 60 km/h, killing 70 people.

Typical mistake

Do not confuse the terms for molten rock. Beneath the Earth's surface it is magma. When it reaches the surface, it is known as lava.

Ashfalls

Tephra emitted during an explosive eruption includes volcanic ash, which consists of fragments of pulverised rock and volcanic glass less than 2 mm in diameter.
- Huge quantities of ash can be dispersed over large areas by the wind.
- Ash is the most frequent and widespread volcanic hazard, but only accounts for 5% of volcano-related deaths, usually due to respiratory conditions.
- It can impact on human activities:
  - The weight of ash collapses buildings. A square metre of ash 10 cm thick can weigh up to 70 kg, or almost double when wet.
  - Interrupts transport by blocking roads and railways and reducing visibility.
  - Clogs engines and machinery. Jet engines can be affected by thinly dispersed ash.

Tephra refers to all explosive eruption materials, including those over 2 mm in diameter. Some descriptions and estimations of the amount of volcanic ash will include it.

- O Destroys and damages crops and farmland. Can poison or result in the starvation of livestock.
- O Wet ash conducts electricity, causing short circuits and failure of high-voltage circuits and transformers.
- The 1991 eruption of Mt Pinatubo in the Philippines had a VEI of 6 and was the second largest eruption of the twentieth century. $10\,km^3$ of rock was blasted up to $40\,km$ in the air, much of it returning as ashfall and adding to the $700 million worth of damage.

Lahars

A lahar is a mixture of volcanic ash, debris and water that travels down a volcanic slope, usually along a valley.
- Water can come from snow and ice melting as a result of an eruption, or from rainfall.
- A lahar's velocity can exceed $200\,km/h$ in steep areas, decelerating and depositing as it moves further away into lowland areas.
- Lahars increase in size as they move downslope.
- They can occur after an eruption, when heavy rain erodes loose volcanic sediment.
- They present the greatest risk to humans after pyroclastic flows. They can engulf settlements, destroy buildings and transport links, and bury agricultural land.
- The 1985 eruption of Nevado del Ruiz in Colombia sent a $60\,km/h$ lahar downslope, engulfing the town of Armero and killing over 20,000 inhabitants.

Jökulhlaups

Jökulhlaups are glacial outburst floods (p. 49).
- Subglacial volcanic activity causes melting at the base of an overlying glacier.
- The trapped meltwater is eventually released when the dam breaks, creating a brief but very significant flood.
- Jökulhlaups tend to occur in sparsely populated areas, limiting the impact they have on humans. They can damage buildings, transport infrastructure and farmland.
- The 1996 eruption of Grimsvotn in Iceland drained $3.2\,km^3$ of water in 40 hours, travelling at 5,000 metres per second.

Volcanic landslides

A volcanic landslide is a mass of volcanic rock or debris that moves rapidly under the force of gravity.
- They are common on volcanic cones where layers of lava and loose rock debris are found on tall, steep slopes.
- They can be triggered by:
  - O volcanic eruptions
  - O earthquakes in the area
  - O heavy rainfall saturating the ground
  - O rising magma causing the ground to deform and resulting in instability
- Landslides can bury areas in debris many metres thick, destroying buildings.
- They can block streams, creating flooded areas.
- They can help create lahars.
- The removal of part of the cone in a landslide may release pressure, which can trigger eruptions.

- The 1980 eruption of Mt St Helens in the USA involved a landslide triggered by an earthquake. $2.9\,km^3$ of material travelled at 170–250 km/h for 20 km, covering an area of $62\,km^2$.

Toxic gases

As magma rises to the surface the pressure drops and gases are released, eventually reaching the atmosphere.

- Harmless water vapour is the most abundant gas, but carbon dioxide, sulfur dioxide and hydrogen sulfide can be emitted, which are all hazardous.
- Carbon dioxide is colourless and odourless and is denser than air, so it sinks and collects in depressions. Humans and animals entering the depression can be asphyxiated.
- Sulfur dioxide can be a skin irritant and causes acid rain. It is also a greenhouse gas, which can play a role in climate change.
- Usually death tolls from toxic gases are low, but the release of carbon dioxide from Lake Nyos in Cameroon in 1986 killed 1,700 people.

Primary impacts of volcanic hazards

Table 5.6 Primary impacts of volcanic hazards

Impact		Scale
Environmental	Reshaping of the landscape by eruption, lahars and landslides	Local/regional
	Destruction of vegetation	Local/regional
	Changes to drainage patterns by lahars and landslides	Local
	Increased precipitation due to released water vapour	Local
	Reduction in temperature through ash clouds reducing incoming solar energy	Global
Demographic/ social	Over 91,000 people were killed by asphyxiation, thermal injuries and trauma from major eruptions between 1900 and 2016; 80% of deaths were from the 10 deadliest eruptions	Regional
	4.7 million people were affected between 1900 and 2016	National
	15,000 were injured between 1900 and 2016	Local

Secondary impacts of volcanic hazards

Table 5.7 Secondary impacts of volcanic hazards

Impact		Scale
Environmental	Emissions of greenhouse gases can add to global warming (p. 92)	Global
	Emission of sulfur dioxide increases the risk of acid rain	Global
Demographic/ social	Average of over 15,000 people left homeless every year	Local
	Poor health from stress-related illness or spread of disease	Local
	Lack of food due to crop damage and lack of clean water	Local
Economic	Loss of income from the destruction of farmland or business; financial loss from associated unemployment	Regional
	Financial loss from loss of primary activities such as agriculture and forestry	Regional/national
	Loss or damage to businesses	Regional/national
	Cost of clean-up operations, repairs and rebuilding, estimated at US$65 million a year	National
	Economic loss from 1900 to 2016 estimated at US$3.5 billion	Global

Contrasting volcanic eruptions

Table 5.8 Comparison of the Eyjafjallajökull and Nevado del Ruiz volcanic eruptions

	Eyjafjallajökull, Iceland	Nevado del Ruiz, Colombia
Date	March and April 2010	November 1985
VEI	4	3
Description of volcano	Composite (stratovolcano) with a large, steep-sided cone 1,651 m high Subglacial	Composite (stratovolcano) 5,389 m high Glacier-covered summit
Eruption	Preceded by earthquakes Ash plume 11 km high Melted and vaporised the glacier above — an estimated 1 km³ of ice in the crater	Ash plume 30 km high 35 million tonnes erupted, creating pyroclastic flows Meltwater from glacier created lahars
Precautions	800 evacuated before eruption Locals given 30 minute warning of floods	Most not evacuated due to communication problems and an inefficient warning system
Deaths	0	23,000
Injured	0	5,000
Homeless	0	8,000
Other effects on local, national and global populations and economic systems	Ash plume blown southeast to Europe 20 countries closed airspace to commercial jet traffic for several days 100,000 flights cancelled, affecting 10 million travellers Airlines lost $1.7 billion in revenue Farmland was flooded and roads and homes damaged by meltwater Water supplies contaminated with fluoride Lack of imported raw materials $100 million economic loss to Iceland $5 billion economic cost to European economy	Lahars engulfed the town of Armero 40 km away, killing 70% of the population 3.4 km² of agricultural land lost and crops destroyed 150,000 animals killed Houses, businesses and infrastructure destroyed $7.7 billion economic cost

Now test yourself

 TESTED ☐

3 What is the difference between an effusive and an explosive eruption?
4 What is the difference between ashfall and a pyroclastic flow?
5 Why are lava flows the least hazardous process of a volcanic eruption?

Answers on pp. 172–3

> **Exam tip**
>
> Apply the information in case studies to the specific question to help make a point. Do not just write a descriptive account of what happened.

Earthquakes, processes, hazards and their impacts

Earthquake characteristics

- Earthquakes are movements of the Earth's crust, usually along pre-existing faults.
- They occur as the result of a gradual build-up of stresses due to crustal movement.
- Rocks that span the margins of adjoining moving plates are subjected to stress and slowly deform.
- When the deformation is too great the rock fractures and returns to its original shape, creating seismic waves — a process known as **elastic rebound**.

Earthquakes produce four main types of seismic wave (Table 5.9).

Table 5.9 Seismic body and surface waves

Body waves	Surface waves
Travel through the interior of the Earth	Travel only through the crust
Higher frequency and arrive before surface waves	Lower frequency and arrive after body waves
P waves (primary):	Responsible for most destruction
<ul><li>Fastest waves and first to arrive</li><li>Can move through solid rocks and liquids</li><li>Caused by compression, pushing and pulling the rock as it moves</li></ul>	L waves (Love):<ul><li>Fastest surface waves</li><li>Move the ground horizontally from side to side</li><li>Often generate the greatest damage</li></ul>
S waves (secondary):<ul><li>Slower than P waves and only move through solid rock</li><li>Move rock particles at right angles to the direction of wave travel</li><li>Second wave to arrive — responsible for much damage</li></ul>	R waves (Rayleigh):<ul><li>Waves follow an elliptical motion, which moves the ground up and down and side to side</li><li>Can be larger than other waves and result in most of the shaking that is felt</li></ul>

Earthquake processes and the production of associated hazards

Ground shaking

- Shaking of the ground caused by the passage of seismic waves represents the greatest hazard to humans due to collapsing structures and the destruction of infrastructure.
- Intensity of the ground shaking depends on:
  - intensity and duration of the earthquake
- distance from the **epicentre**
  - geology — solid bedrock is less subject to intense shaking than loose sediment

> The **epicentre** is the point on the Earth's surface directly above the focus.

Liquefaction

- Loose sediments such as sand and silt combine with groundwater and behave like a fluid or quicksand.
- It can cause the foundations of buildings to sink, causing them to collapse.
- After the earthquake the water sinks deeper into the ground and the surface firms.

Landslides

- Earthquakes create stresses that make weak slopes fail, triggering landslides as well as rockfalls and avalanches.
- Landslides can significantly increase the death toll from an earthquake and hamper rescue efforts.
- In 1970 an earthquake 25 km off the coast of Peru triggered a landslide on Mt Huascaran 130 km away. Travelling at over 200 km/h, it added 20,000 people to the death toll.

> The **focus** is the point where the rocks fracture. It can be up to 700 km below the surface. It is also called the hypocentre.

Tsunamis

- Caused by a shallow **focus**, undersea earthquake at a plate margin, creating a sudden rise or fall in the ocean floor, which displaces the water above it (Figure 5.11).
- Ocean waves spread out from the epicentre, travelling at hundreds of kilometres per hour.
- Wave height is low in the ocean but increases to up to 30 m as the water depth decreases near land.
- Tsunamis can travel over 10 km inland, causing destruction, flooding and carrying debris.
- Over 80% of tsunamis occur within the Pacific Ring of Fire.
- Tsunamis are the deadliest secondary effect of earthquakes. The ten most devastating tsunamis since 1960 have killed 271,700 people, 85% in the 2004 Indian Ocean tsunami.

> **Exam tip**
>
> The impact of a tsunami is an example of synoptic links between different components of the specification. As well as its impact on humans as a secondary tectonic hazard, it will also impact on coastal systems. For example, it could be seen as a cause of metastable equilibrium (p. 8)

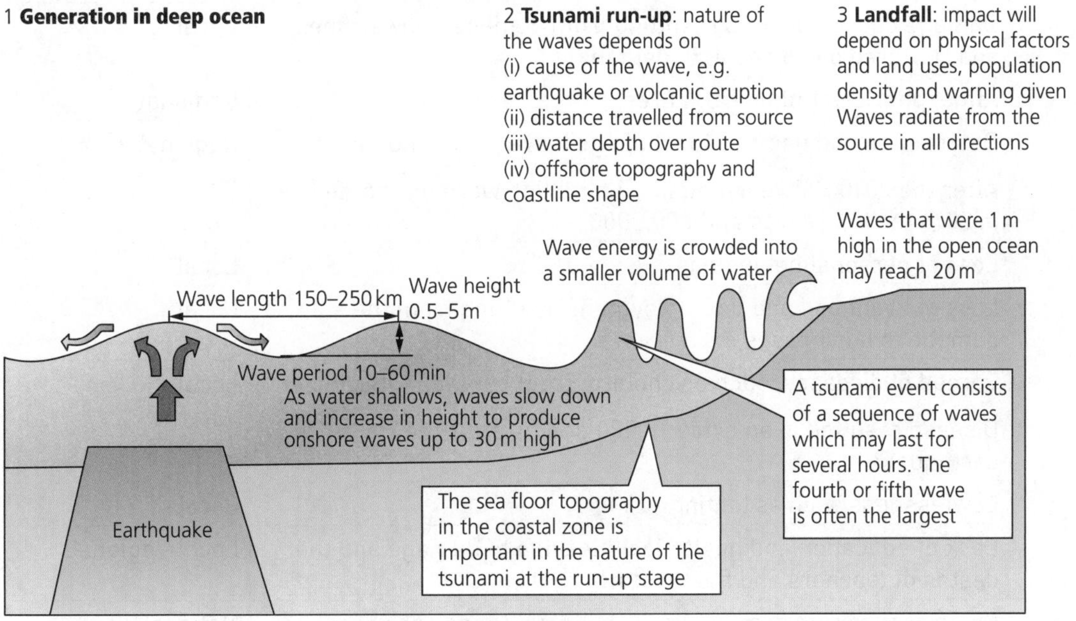

1 **Generation in deep ocean**

2 **Tsunami run-up**: nature of the waves depends on (i) cause of the wave, e.g. earthquake or volcanic eruption (ii) distance travelled from source (iii) water depth over route (iv) offshore topography and coastline shape

3 **Landfall**: impact will depend on physical factors and land uses, population density and warning given Waves radiate from the source in all directions

Figure 5.11 Formation and features of a tsunami

Primary impacts of earthquakes

Table 5.10 Primary impacts of earthquakes

Impact		Scale
Environmental	Landslides, liquefaction and tsunamis	Local/regional
	Land uplift or subsidence In the 1964 Alaskan earthquake some areas were permanently raised by 9 m while others dropped by 2.4 m, resulting in flooding	Local/regional
	Damage and destruction to buildings, roads, railways, power supplies, and water and sewerage systems	Local
Demographic/ social	Earthquakes killed 804,000 people between 2000 and 2017, mainly due to collapsed structures	Local
	Many people injured — injuries usually outnumber deaths The deadliest earthquake in 2017, in Iran, killed 630 but injured 8,485 people	Local

Secondary impacts of earthquakes

Table 5.11 Secondary impacts of earthquakes

Impact		Scale
Environmental	Damage and flooding resulting from the primary impacts	Local/regional
	Damage caused by fires resulting from damaged power lines and ruptured gas pipes and fuel tanks	Local
	Atmospheric pollution from fires	Regional
Demographic/ social	Destruction and damage to buildings leave people homeless After the 2010 Haiti earthquake, 1.5 million were left homeless; in 2016 the figure was still 600,000	Regional
	Loss of personal belongings	Local
	Loss of livelihood and unemployment resulting from the damage to businesses	Local
	Spread of diseases, such as cholera, from contaminated water Disease resulted in an extra 10,000 deaths after the Haiti earthquake	Local
	Lawlessness such as looting	Local
	Lack of education and health facilities due to damage and the deaths of teachers and health staff	Local/regional
Economic	Costs of repairing damage and rebuilding; these can vary greatly depending on location of the earthquake	National
	Loss of production and economic activity due to damaged buildings and transport difficulties	National
	Economic damage from a large earthquake in a developed country between 1985 and 2015 varied from $2 million to $232 billion (2015 equivalent prices)	National

Contrasting earthquakes

Table 5.12 Comparison of the Bam and Christchurch earthquakes

	Bam, Iran	Christchurch, New Zealand
Date	26 December 2003	22 February 2011
Magnitude: Richter/Mercalli	6.8/IX (violent)	6.2/VIII (destructive)
Cause	Movement of the Arabian plate against the Eurasian plate at 3 cm a year	Rupture along a 'hidden' fault line Possibly an aftershock from the 7.3 Canterbury, New Zealand earthquake of 2010
Precautions	Little earthquake education	Earthquake education Awareness of previous earthquake Some earthquake-resistant buildings
Deaths	26,271	185
Injured	Over 20,000	220 major; 6,800 minor
Homeless	75,000	? (10,000 buildings to be demolished)
Other effects	85% of buildings destroyed; mud brick buildings collapsed; in many cases building regulations had not been followed Electricity, water supply and sanitation services destroyed or affected Health system destroyed and many health staff killed Education system destroyed and 5,400 teachers killed Economic cost was $1.9 billion; $30 million was needed to restore the health service Over 40 countries sent aid Freezing temperatures helped prevent the spread of disease after the event	Liquefaction and shaking damaged 100,000 buildings; some collapsed (buildings may have been weakened by the 2010 Canterbury earthquake) 10,000 buildings had to be demolished Water supply and sewerage systems damaged Some buildings flooded Rescue crews sent from many other countries Some areas of Christchurch abandoned due to the risk of liquefaction One-fifth of the population left Christchurch permanently Economic cost was $11 billion

Now test yourself
TESTED

6 Apart from magnitude, why do some earthquakes have a much greater impact than others?

Answers on p. 173

Human factors affecting risk and vulnerability
REVISED

- **Risk** and **vulnerability** are important factors influencing the impact of a tectonic **hazard**.
- They are major factors in a hazard event becoming a **disaster**.
- The level of hazard risk can be measured with the following risk equation:

$$\text{risk} = \frac{\text{hazard (frequency and/or magnitude)} \times \text{level of vulnerability}}{\text{resilience level (ability of community to cope)}}$$

Economic factors

- The level of poverty of countries and individuals increases the degree of vulnerability. Poverty results in:
  - a lack of education, resulting in lack of knowledge of the risks and hazard perception
  - poor-quality building standards, increasing the risk of collapse
  - limited ability to cope with the aftermath of the event
  - a lack of alternatives to reduce the risk, such as moving to a safer area
  - a lack of technology to monitor hazards and issue warnings
- Wealthy countries can afford technology to monitor tectonic areas (especially volcanoes), establish warning systems (e.g. Pacific tsunami warnings), and mitigate the impacts (e.g. with earthquake-resistant buildings).
- Wealthy people may have a greater ability to evacuate the hazard zone, for example driving out of a tsunami zone after a warning.
- Economic benefits such as high crop yields from fertile land may encourage people to stay in the hazard area.
- Poor people are less likely to have insurance to help cover losses.
- Wealthy areas and individuals are more vulnerable to economic loss due to destruction of and damage to higher-value buildings, infrastructure and businesses.
- Reliance on technology for everyday life is greater in wealthy countries, making them more vulnerable to problems if the hazard affects the technology systems, for example damage to seafloor cables (p. 123).

Social factors

- The greater the number of people living in a hazard zone, the greater the potential for loss of life or injury, or economic loss. For example, the 2001 Kunlun earthquake in China was magnitude 8 but caused no loss of life because the area was remote and uninhabited.
- The young and elderly may be more vulnerable because they have less physical strength to survive disasters and may be more susceptible to diseases. They may also be dependent upon others for survival.
- Women may be more vulnerable in many societies due to a lack of education, fewer financial resources and the role of protecting other family members. Gender inequalities can increase women's vulnerability, for example in some parts of the world, such as Afghanistan and Yemen, a woman is not allowed to leave the house if her husband chooses to prevent her from doing so.
- Poorly educated people may be unaware of the potential risks and how to mitigate them. For example, people unaware of the warning signs prior to the Indian Ocean tsunami in 2004 walked out towards the sea to see what was happening.
- The impact can be influenced by a person's or community's perception of the hazard, such as:
  - acceptance that it might happen and that they will bear the loss
  - complacency due to false alarms from warning systems
  - fear of the unknown — is leaving better than staying, or is it better to stay and protect property, for example from looting?

Risk is the probability of a hazard occurring and causing loss of life and livelihoods.

Vulnerability refers to the ability of a community to cope with the impacts of a hazard.

A **hazard** is a natural event that threatens or causes injury, death and/or damage.

A **disaster** is the result of a hazard seriously disrupting the functioning of a vulnerable community, causing loss of life and material and economic damage.

Political factors

Poor-quality governance can increase vulnerability, for example:

- a lack of development or enforcement of building codes
- bureaucracy and corruption, which may prevent aid from quickly reaching disaster areas
- a lack of hazard mitigation planning

In the Nepal earthquake of 2015 8,700 died and 500,000 homes were destroyed as building standards had been ignored, often with extra floors being added as money allowed. The area had 12 fire engines, which were mostly old models and out of service. International aid donors gave money to NGOs rather than the Nepalese government.

Geographical factors

- Urban areas with a greater population density are more vulnerable to loss than rural areas.
- Rural areas may be remote and so more vulnerable due to the difficulties of rescue and aid reaching the area. For example, in the Nepal earthquake, hillside villages were completely cut off as roads were destroyed, delaying the arrival of aid.
- The time the hazard occurs can increase the risk, for example the 2003 Bam earthquake in Iran (Table 5.12) occurred at 5:26 a.m., when many people were inside and so were crushed while sleeping.
- Vulnerability may increase due to the risk of other hazards in the area, for example the impact of the eruption of Mt Pinatubo in the Philippines in 1991 was exacerbated by the arrival of Typhoon Yunya, adding to the lahars and increasing the weight of ash deposits, leading to structural damage.

The **pressure and release (PAR) model** (Figure 5.12) shows how the level of impact is due to the level of vulnerability, which in turn is influenced by the socio–economic characteristics of the area.

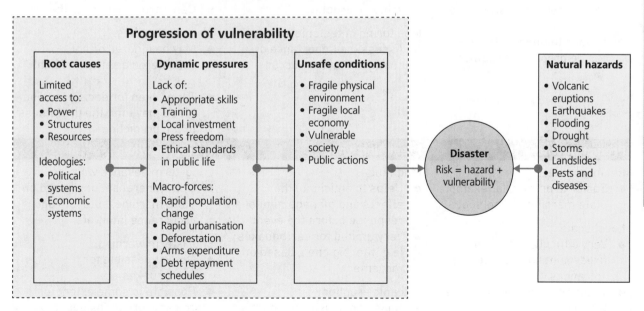

Figure 5.12 The pressure and release model

Now test yourself

7 What is the difference between a hazard event and a disaster?
8 How can good governance decrease the hazard vulnerability of a community?

Answers on p. 173

Responses to tectonic hazards

Responses can involve:
- monitoring and attempting to predict and warn of a hazard event
- mitigating the hazard by modifications

Monitoring, prediction and warning of tectonic hazards

Attempts to reduce a community's vulnerability to a hazard could involve:
- evacuation to areas of safety
- preparations for the event
- risk assessment by authorities and insurance companies
- allowing providers of post-event aid to be prepared

Table 5.13 **Factors influencing tectonic hazard prediction**

Why predict?	Evidence base for prediction	Who is prediction for?
Reduce impacts: • Evacuate and prepare, to reduce the death toll • Development of management plans • Risk assessments and cost-benefit analysis can be carried out • Actions to mitigate the event can be prepared by authorities	Past records and recurrence interval: • Allows assessment of the likelihood of an event • Allows assessment of the type of event, such as what type of volcanic eruption Monitoring of tectonic activity • Assess the importance of changes that may occur	Warn the population: • May save lives if correct • Could make the situation worse by creating panic • Complacency could set in if continually over-warned • Can attract visitors to the area Warn authorities: • May be ignored or not considered a serious enough risk • Time taken for decisions to be made may limit the usefulness of the prediction
Predicting where	**Predicting what**	**Predicting when**
Regional scale: • Easier to predict, for example plate margins and hotspots Local scale: • Very difficult, except for fixed-point hazards such as volcanoes • Tsunamis' movement can be tracked and predicted	Magnitude: • Helps to anticipate the effects and management of a response before the event • Very limited for earthquakes (e.g. the 'big one'), based on movement Volcanic eruption: • Based on activity and past eruptions Potential impacts: • Aids the organisation of a response to the event, for example the type of aid required	Long-term advance warning: • Only generalisations based on plate margins, hotspots and recurrence intervals Short-term warning: • Not yet possible for earthquakes • Possible to some extent for volcanic eruptions due to increasing tectonic activity • Tracking of tsunamis allows the estimation of the time of arrival

Earthquakes

Scientists use a variety of monitoring methods to investigate the possibility of earthquake prediction (Table 5.14).

Table 5.14 Monitoring methods to aid earthquake prediction

Method	Reasons for monitoring
Laser reflector	Monitors small movements along a fault
Creep meter	
Gravity meter	
Tilt meter	
Strain meter	Measures changes in the stress in rocks
Well levels	Monitors groundwater movements and height of the water table, which can change just before an earthquake
Groundwater measurements	
Radon gas meter	Levels of radon gas dissolved in water can increase before an earthquake
Magnetometer	Changes in the Earth's magnetic field have been recorded before an earthquake
Seismograph	Recording smaller foreshocks
Animal behaviour	Anecdotal evidence abounds, for example rats and snakes leaving homes and heading for safety before an earthquake

- The United States Geological Survey (USGS) suggests that an earthquake prediction must aim to define three elements:
  - the date and time
  - the location
  - the magnitude
- No accurate prediction has yet been made that meets all three elements.
- Unofficial predictions (e.g. on social media) can occur when something happens that is thought to be a precursor to a major quake, such as swarms of small quakes, changing levels of radon in water or strange animal behaviour.
- Many such activities frequently occur without an earthquake following.
- The 1975 Haicheng earthquake in China was forecast based on small earthquakes and animal behaviour. People chose to sleep outside and were saved when an earthquake caused destruction.
- No prediction was made of the Great Tangshan earthquake the following year, when thousands died.
- A number of countries, such as Japan and Mexico, have started using **earthquake early warning (EEW)** systems, which give regional warnings seconds or minutes before the event.

Volcanoes

- Many eruptions are preceded by environmental changes. Monitoring changes in volcanoes can aid in the prediction of a possible eruption (Table 5.15).

Table 5.15 Monitoring methods to aid volcanic eruption prediction

Environmental change	Reasons for monitoring
Seismic activity	Seismic data over time may show an increase in activity
	If earthquakes are migrating towards the surface, this could suggest rising magma fracturing rocks
Ground deformation	Tilt meters and laser measures can detect minute changes in slope angle and the distance between points, suggesting that rising magma is displacing the ground above
Gas emissions	Ground-based, airborne and satellite detectors can measure changes in the amount and composition of gas emissions
	If the gases pass through lakes, changes in colour and acidity of the water can be recorded
Thermal changes	Detectors, including satellites, can register changes in temperature at the surface and in lakes, suggesting an increase in activity
	Increasing discharge from hot springs and death of vegetation are visible signs
Lahar monitoring	Observations of lahars, including by remote video cams, drones and seismometers recording vibrations from an active lahar, can allow short-term warnings

- While monitoring evidence can predict the increased likelihood of an eruption, it cannot give an accurate prediction that meets all three of the USGS requirements. However, warnings can allow plans to be prepared.
- Monitoring of Mt Pinatubo in 1991 resulted in the most hazardous zone being officially evacuated over 2 months before the major eruption. A larger area was evacuated just 2 days before, after further monitoring suggested that a major eruption was imminent.

Tsunamis

- Scientists can only predict the possibility of a tsunami by monitoring undersea seismic activity.
- Tide gauges, tsunami detection buoys and pressure recorders located on the ocean bottom, are used to record changes that show a tsunami has been generated.
- Once a tsunami is generated, areas can be warned of its arrival. The Pacific tsunami warning system coordinates tsunami threat information throughout the Pacific and has been issuing warnings for 70 years, allowing evasive action to take place.

Mitigating tectonic hazards

Reducing the impact of tectonic hazards can be approached in three different ways (Table 5.16):
- Modify the hazard event.
- Modify human vulnerability to the event.
- Modify the loss from the event.

Table 5.16 Methods of tectonic hazard mitigation

Tectonic hazard	Modify the event	Modify human vulnerability	Modify the loss
Volcanic eruption	Control the flow of lava, for example: • 7 billion litres of sea water sprayed on to advancing lava to cool and solidify it before it reached Vestmannaeyjar harbour in Iceland in 1973 • Barriers diverting lava flows away from villages on Mt Etna in Sicily in 1991 Both methods possibly only worked because the eruption ceased	Monitoring and prediction Warning systems Evacuation plans Hazard mapping, for example of lahar or lava flow risks, which can influence land use planning and zoning — some limited use in Hawaii and New Zealand Education to help a community learn how to prepare for an event and to reduce vulnerability	Rescue and relief efforts Aid from governments and NGOs to help with rescue as well as with food, clean water and shelter Insurance, however: • mainly available in economically wealthy nations • due to the risk factor, the cost for individuals may be too high and outweigh the benefits — it may require households to take many preventative measures Individual responses — individuals may adopt changes to help modify losses in a future hazard event, for example purchase of emergency earthquake survival kits
Earthquake	Not yet possible	Building codes requiring aseismic (earthquake-resistant) buildings, such as in Japan and USA. However: • difficult to retrofit to old buildings • costs are prohibitive for most buildings except important public buildings • in many developing countries it may be difficult to enforce the codes due to lack of political will or money Education and earthquake drills to reduce vulnerability Hazard mapping, for example of areas prone to liquefaction to prevent building in the area, as in the planning of the rebuilding of Christchurch, New Zealand Increasing use of smart technology, which sends short-term warnings to devices, allowing people to take immediate action	
Tsunamis	Coastal defences and hard engineering, for example Japan has built a 400 km chain of concrete sea walls, in places 12 m high Soft engineering, for example the redevelopment of mangroves to protect rural areas by dissipating the wave energy (p. 22)	Warning and prediction systems Education Clearly marked evacuation routes Coastal zone management and land use planning, for example set-back zones, where development is not allowed	

Short-term and long-term responses to the effects of earthquake and volcanic hazards

- The response to a tectonic hazard depends on many interrelated physical and human factors.
- The range of factors can be seen in the **hazard disaster management cycle** (Figure 5.13).

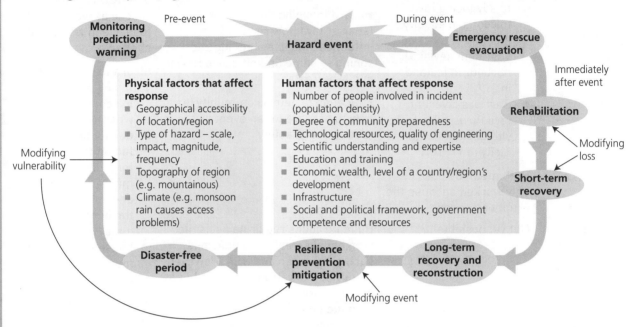

Figure 5.13 The hazard disaster management cycle

- The interaction of factors will influence the response in the short term, during and immediately after the event, and in the longer term, where future strategies to modify an event and the level of vulnerability may influence the response.
- Park's disaster–response curve (Figure 5.14) shows how a hazard event and subsequent responses impact on quality of life.
- Stage 5 shows the long-term response, with quality of life being restored or even improved if vulnerability has been reduced.
- The model can be used to compare long-term responses to different hazard events.

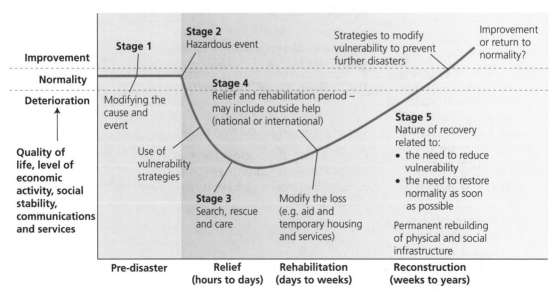

Figure 5.14 Park's disaster–response curve

Exam practice answers and quick quizzes at www.hoddereducation.co.uk/myrevisionnotesdownloads

Exam practice

The format for the different examination papers is shown below.

Specification	Method of examination
Eduqas A-level Component 3; Section A	One compulsory extended-response question
Eduqas AS Component 1; Section B	One compulsory structured data-response question and two extended-response questions
WJEC AS Unit 1; Section A	Three compulsory data-response questions
WJEC A2 Unit 4; Section A	One compulsory extended-response question

Eduqas A-level format

1 'The economic development of an area is the most significant factor determining the impact of tectonic hazards.' Discuss. [38]

WJEC A2 format

2 Discuss the role social factors play in determining the impact of tectonic hazards. [20]

WJEC AS format

3 Study Figure 1, which shows new volcanic eruptions between 1950 and 2015.

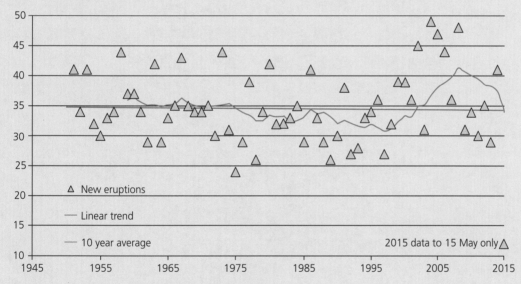

Figure 1 New volcanic eruptions 1950–2015

a) Describe the trends in volcanic activity shown in Figure 1. [5]
b) Outline the limitations of the data shown in Figure 1. [5]
c) Suggest reasons why some communities may be more at risk than others from volcanic activity. [10]

Answers and quick quiz 5 online

ONLINE

Summary

You should now have an understanding of the following:

- The Earth comprises a number of layers — inner and outer core, mantle and crust — which have different physical properties.
- The Earth's surface is made up of rigid plates that move in relation to each other.
- The main zones of earthquakes and volcanoes follow plate boundaries. Volcanoes are not found at conservative plate margins but are found at hotspots away from margins.
- Volcanoes can be classified by shape and type of eruption.
- Earthquakes result from movements along fractures and faults in rocks.
- Tectonic hazards can have a significant impact on the environment.

- The demographic, social and economic impacts of tectonic hazards depend on physical factors, such as magnitude, and human factors, such as population density and level of development.
- Economic, social, political and geographical factors can influence the levels of risk from, and vulnerability to, a tectonic hazard.
- Vulnerability can be modified by monitoring, predicting and warning of tectonic hazards. Predicting earthquakes is not yet possible.
- The choice of response to a tectonic hazard depends on many interrelated physical and human factors.

Exam practice answers and quick quizzes at **www.hoddereducation.co.uk/myrevisionnotesdownloads**

Now test yourself answers

Chapter 1

1 The coast is an open system because energy and sediment are able to move from the system to the surrounding environment.

2 A sediment cell is a self-contained system of sediment flows and stores found on a particular stretch of coastline.

3 Some coastlines are experiencing changes caused by processes operating over a very long period (lasting millennia), such as a post-glacial eustatic rise in sea level.

4 Positive feedback increases the impact of the initial change that occurred to a system, whereas negative feedback reduces the effect of the initial change.

5 – Wind strength — stronger winds produce bigger waves.
  – Fetch — the longer the uninterrupted distance the wind blows over the water, the bigger the waves.
  – Wind duration — a longer duration creates bigger waves.
  – Water depth — deeper water reduces the wave amplitude.

6 A high-energy coastal environment mainly features landforms characterised by erosion, such as cliffs and wave-cut platforms, whereas a low-energy coastal environment is characterised by depositional features such as sand dunes, beaches and mudflats.

7 – Geology — whether the rock is hard or soft.
  – Rock structure — the numbers of joints and bedding planes influence the impact of erosional processes.
  – Climate — influences the type and rate of weathering.
  – Wave type — influences whether erosion or deposition is the dominant process. Impacts on the rate of erosion.
  – Shape of the coastline — wave refraction can concentrate wave energy at headlands, causing erosion and resulting in deposition in bays.
  – Human activity — can influence longshore drift, which can impact on depositional features. Management strategies can impact on erosion rates.

8 Sub-aerial processes are weathering and mass movement, which act upon the coast, especially cliff faces.

9 Marine erosion processes, especially hydraulic action and wave quarrying, exploit lines of weakness, creating caves either side of a headland. Continued erosion results in the caves joining, forming an arch. Enlargement of the arch results in rockfall as the roof collapses, leaving a stack. Sub-aerial processes such as physical weathering and mass movement impact upon the feature at all stages of its development.

10 The coastline will adjust to the reduced input of sediment to reach a new state of equilibrium. With less sediment deposited to maintain the features, erosional processes may have more of an impact, reducing the size of the landform, with eroded material possibly transported further along the coast by longshore drift.

11 Vegetation reduces wind velocity, which causes sand to be deposited and accumulate. Plants such as marram grass continue to grow as the sand accumulates, trapping more sand and encouraging a dune to form. Plant roots can help bind the sand, reducing the possibility of wind erosion. Decaying vegetation encourages soil formation and further vegetation cover, protecting the dunes from erosion by the wind.

12 a) Eustatic refers to the global change in the volume of water in the oceans, for example due to melting of continental ice sheets.

   b) Isostatic refers to a localised change in the relative sea level as a result of the downward or upward movement of land masses, for example due to the weight of ice being added or removed during and after glacial periods. This is typically a very slow process of change, lasting many millennia.

13 The movement of sediment is self-contained within the cell. Any management strategy that impacts on the input or transport of sediment is likely to have consequences for the processes and landforms at other interconnected parts of the coastline within the cell.

14 A glacier is an open system because water and sediment can move from the system to the surrounding environment.

15 The glacial budget is the balance between inputs in the accumulation zone and outputs in the ablation zone.

16 Positive feedback increases the impact of an initial change that occurred in a system, whereas negative feedback reduces the effect of the initial change.

17 Pressure melting point is the temperature at which ice melts due to pressure from the weight of ice. The temperature can be below 0°C.

18 Cold-based glaciers are slower moving. Most movement is by internal deformation — intergranular flow and laminar flow. Warm-based glaciers are usually faster. They move by basal sliding — enhanced basal creep and regelation slip — and sub-glacial bed deformation. They can also move by the same methods as cold-based glaciers.

19 – Gradient — steeper slopes cause faster movement.
– Altitude — influences the temperature, which in turn controls the mass balance by influencing accumulation and the amount of meltwater.
– Size — a greater mass has greater potential velocity.
– Lithology — influences the level of friction at the glacier base.
– A build-up of meltwater can create surge conditions.

20 – Warm-based glaciers and glaciers with faster velocities are more actively moving and have more erosion potential.
– Ice thickness — thicker ice exerts greater pressure, increasing erosion potential.
– Bedrock — well-jointed bedrock allows plucking to occur more readily. Permeability influences the amount of meltwater, which can affect the level of plucking and sub-glacial fluvial erosion.
– Debris — large quantities of angular debris can increase abrasion at the glacier base.

21 Freeze–thaw weathering and transportation by meltwater create a nivation hollow. Abrasion and plucking occur as the glacier moves in a rotational manner. The glacier transports eroded material.

22 A rôche moutonnée is formed by the ice flowing over a resistant rock outcrop, eroding the up-valley slope smooth by abrasion and the down-valley slope jagged by plucking. A crag and tail is formed by ice flowing around a rock outcrop, resulting in a steep up-valley slope and a gently sloping down-valley slope.

23 A terminal moraine is a ridge across a valley deposited at the point of maximum advance of a glacier. A push moraine is a ridge of previously deposited material pushed into place when a glacier re-advances.

24 An ice-contact deposit is material deposited by sub-glacial meltwater, forming features such as eskers. A proglacial deposit is material deposited down-valley from the glacier, forming features such as sandurs (outwash plains).

25 A relict feature is landform that was formed under past climatic conditions and by processes that are very different from those found today.

26 Periglacial refers to places at the edges of glacial areas, where seasonal freezing and thawing modify the landscape.

Chapter 2

1 The character of a place can be influenced by many factors, including population size and structure, socio-economic factors such as employment rates, income and crime rates, cultural ideas of the population, the location and type of environment, and the actions of authorities. Flows of people such as migrants, investments and resources can also shape the character.

2 Perception can be influenced by a person's age, gender, socio-economic status and socio-cultural position, as well as (formal and informal) information from direct and indirect experience of the place.

3 Globalisation is the integration and interaction of economies and societies by a global network of trade, communication and immigration.

4 Deindustrialisation is the reduction of industrial employment and/or output resulting in economic and social change and challenges for places. Sometimes employment may fall but output stays the same, for example as a result of automation.

5 The idea that poverty resulting in social inequalities such as poor living conditions, ill health and poor skills maintains or even amplifies (via positive feedback) the level of poverty in a place.

6 Deindustrialisation causes a decrease in employment opportunities. Consequently, the affected population has less disposable income to spend. Shops and services are used less, making less money, and close as they are no longer economically viable, increasing the initial impact. An example of positive feedback.

7 Re-urbanisation involves the movement of people back into city centres. Many who move back are high-income, young professionals who can afford to renovate older housing. Older industrial buildings are converted into luxury accommodation for the new population and services develop in the area to cater for their tastes. The area has become gentrified, changing from a poor area to a richer one.

8 Clusters can attract a higher-educated, digitally proficient workforce, and may have links to universities. Cluster areas may receive support from the government and can attract investment from MNCs. They can benefit from highly developed infrastructure and relaxed planning

controls. Over time, clusters can gain worldwide influence, becoming global hubs.

9 Non-quaternary workers may not be able to afford higher housing costs created by the demand from higher-paid workers. The workers attracted by the cluster may change the social character of a place, and services cater for them. Non-quaternary workers may suffer social exclusion as a result, as well as the impacts of a lack of suitable job opportunities.

10 The rural area may increase in popularity, increasing visitor numbers and resulting in traffic congestion and environmental damage, which local people may resent. Demand to move into the area can raise house prices out of reach of locals (rural gentrification). Second home ownership and changes to services to cater for visitors may change the character of an area, creating local resentment.

11 The area may become a desirable place to live, increasing house prices and excluding many of the local population. Increases in visitor numbers can cause issues with parking, congestion and noise that affect residents (overheating). The area may attract all the available investment, leading to decline in other parts of the urban area.

Chapter 3

1 The physical inputs and outputs of matter (water) come from stores within the system. None moves into or out of the system by crossing boundaries.

2 97% of all the Earth's water is salt water and so of little use. 68.7% of the remaining water is unavailable because it is stored in the cryosphere in the form of ice. Small quantities in the atmosphere and biosphere are also unavailable for human use.

3 a) Geology — the permeability of rocks controls the amount of percolation and groundwater storage, which influences the rate of runoff.

   b) Dam construction — creates reservoirs where water can be stored and the flow of water controlled and evened out throughout the year.

4 Saturation excess overland flow occurs when the soil is saturated, so no more rainfall can infiltrate, whereas infiltration excess is the result of rainfall intensity being greater than the infiltration rate.

5 The hydrograph has a steep rising limb, a short lag time, high peak discharge and a steep falling limb.

6 Hydrological drought is a shortage in surface and groundwater stores that may result in not enough water being available to meet human demand, creating the need for rationing. Agricultural drought occurs when there is insufficient moisture to maintain crop yields, which could result in less food being produced to meet human needs and higher prices. Meteorological drought may have less impact because sufficient water may be stored for use or it may occur when it is not the crop-growing season.

7 99.9% of global carbon is stored in the lithosphere and sedimentary rocks. The rest is found in the oceans, fossil fuels, biosphere and atmosphere.

8 It is a slow cycle because it takes a very long time. The average residence time on land is 150 million years. Rock particles have to be weathered and eroded, and transported to the oceans. Sequestration by sediments being deposited, forming new pools in sedimentary rocks, and then earth movements to uplift the rocks above sea level for the cycle to restart, happens over millennia.

9 The rate of primary productivity in a biome will vary as a result of the climatic conditions. Biomes with larger amounts of biomass have a larger carbon store. Where biomes have a more complex structure, for example many layers (e.g. understorey, shrub layer, canopy layer), they have the potential to store more carbon.

10 Population growth increased the use of fossil fuels and the rate of deforestation at a global scale, decreasing carbon storage in the lithosphere (coal, oil) and biosphere, thereby increasing CO_2 emissions.

11 Land and sea areas formerly covered in ice now have a lower albedo and absorb more incoming solar energy, warming the surface. Heat radiating from the surface warms the air, causing the atmospheric temperature to rise. This causes positive feedback as higher temperatures lead to more ice melting. Warmer temperatures may increase ground thawing and decomposition in tundra biomes, releasing more methane and CO_2 (both greenhouse gases). Warmer oceans release more CO_2, contributing to further global warming.

12 Interactions and feedback loops in the systems are complex, making it difficult to predict the full consequences. Technological and economic developments may positively or negatively influence the impact of any changes within the cycles.

Chapter 4

1 Improvements in transport and communication technology mean that goods, data and people can move around the world at a faster rate, reducing the influence and perception of distance.

2 Reduced travel times mean that distance is less of an obstacle to migration. The ability to be able to communicate easily with other parts of the globe reduces the isolation of migrants when they move. Social media allow potential migrants to see the advantages of moving and can provide support, encouraging migration.

3 Diaspora communities can provide support to migrants when they arrive, and may offer employment opportunities. They can also cater for the cultural needs of the migrant. This may make the experience of migrating easier.

4 Immigrants may fill labour shortages in both skilled and unskilled work. A country may suffer from an ageing population and need youthful workers.

5 Hubs can encourage investment from MNCs as well as global institutions. This may increase employment opportunities for the population and boost economic growth in the area.

6 Remittances can alleviate poverty and lead to an increase in spending, which creates a positive multiplier effect in the source country.

7 A global economic recession (e.g. 2007–09) can result in the cancellation of projects or reduced demand, creating unemployment for migrants and a reduction in remittances to the source country. Countries manufacturing components or supplying raw materials may suffer from declining demand. Migrants may be forced to leave or move to hubs, creating labour shortages in host areas. There are also environmental risks, such as climate change and oceanic pollution.

8 Refugees have been forced to leave their home region due to persecution, conflict, economic actions or natural disasters, whereas economic migrants move voluntarily in search of work and an improved quality of life.

9 Refugees may fill labour shortages, including in skilled jobs. Their existing skills can save the host nation the costs of training new workers. They may establish more businesses, creating employment.

10 Outsourcing involves a business hiring foreign companies to produce goods or provide a service, whereas offshoring involves a business moving its own factories and offices abroad to take advantage of factors such as lower costs.

11 EPZs offer many employment opportunities, which encourage people to move to the area. Incomes may be higher than in rural areas. In some parts of the world many jobs are suitable for young, unmarried, poorly educated females whose job opportunities in rural areas are limited.

12 Top-down developments are planned and carried out by authorities, whereas bottom-up involve people and communities making their own developments.

13 Passession allows a nation to claim the EEZ around the island, giving ownership of resources 200 km (and sometimes up to 370 km) from the shore.

14 53% of oil travels through chokepoints. Secure control avoids them being blocked, reducing the supply rate and increasing transport costs, which would increase oil prices and have economic consequences. Tankers are less vulnerable to attack or accidents (with possible environmental consequences).

15 Containers allow large boats to be used that can carry more goods easily. They can be loaded and unloaded quickly without being opened and can be easily transferred between different methods of transport, so that goods spend less time in transit.

16 Most are owned by MNCs such as Google, which can afford the high costs involved, especially when working in a consortium. MNCs want access to their own network to meet their own secure data requirements.

17 The nations claim that the continental shelf is part of their land because it is part of the continental crust. Ownership would give rights to ocean resources and extend the nations' EEZs.

18 Quotas limit the number of fish that can be caught, so that fish stocks can recover or are maintained, allowing the fish to be available for future generations.

19 Oil tankers must have a double hull to prevent spillages in the event of an accident. Empty tanks cannot be flushed with sea water, which is then emptied into the sea. Water used as ballast must be in separate tanks and not those used for oil.

Chapter 5

1 There are more earthquakes, and earthquake zones are much larger than areas of volcanoes. Many are located on the continents, in densely populated areas.

2 The profile only shows the physical features of the event. It does not show population density, level of development and vulnerability of the population in the affected area, which will influence the impact.

3 In an effusive eruption dissolved gases escape easily, allowing runny lava to flow easily downhill. In an explosive eruption pressure builds up until gas explosions blast rock into the air. Lava flows are thicker and slower moving.

4 Ashfall involves volcanic ash, which has been emitted during an explosive eruption, falling back to earth, unlike a pyroclastic flow, where volcanic fragments and ash mixed with gas moves down the sides of a volcano at very high speed.

5 Lava flows tend to only destroy buildings and infrastructure. They can be slow moving, usually giving people time to move out of danger. It is sometimes possible to predict where the lava flow will go.

6 – The location of the epicentre — there may be greater impact in densely populated areas.
 – Vulnerability of the population — there will be a greater impact in areas where hazard perception is weak, building construction is not earthquake-resistant and there is limited ability to cope with the effects.

 – An earthquake can result in a tsunami, landslide, liquefaction or flooding, increasing the impact on people. The time an earthquake occurs and the season may affect where people are located, changing the vulnerability.

7 A hazard is a natural event that threatens or causes injury, death and damage. When a hazard seriously disrupts the functioning of a community through death and damage, it becomes a disaster.

8 Good governance should use its powers to minimise the impact of a hazard. This could be by hazard education to reduce vulnerability, development and enforcement of building codes, land use zoning and emergency planning (e.g. evacuation plans).

Specialised concepts

An outcome of studying geography is gaining knowledge of specialised concepts. These should be used whenever possible because the examiners will expect you to understand what they mean.

adaptation: the ability to adjust to events or to changes in environmental conditions.

attachment: the linkages between individuals, communities and places.

causality: the relationship between cause and effect. It is what connects one process (the cause) with another process (the effect).

equilibrium: the condition of a system when inputs and outputs are balanced.

feedback: the process by which a system responds to changes within the human or physical system. Positive feedback accelerates change. Negative feedback reduces the impact.

globalisation: the process of economies and societies becoming integrated by a global network of trade, communication and immigration.

identity: how individuals view changing places from different perspectives and experiences. The actual or perceived characteristics of a place that make it distinct.

inequality: the extreme social and/or economic differences between people and between places.

interdependence: the mutual reliance of two or more things on each other. This can refer to countries, communities, human activities or parts of the physical environment.

mitigation: the elimination or reduction of an event or process that has a negative effect on humans or the environment.

place: an area of geographic space to which meaning has been given as a result of its identity, characteristics or uniqueness.

representation: the way a place is portrayed either formally or informally. How people are represented in a political sense.

resilience: the ability to recover from events that have a negative impact.

risk: the probability of a negative outcome as a result of physical processes or human activity.

scale: the extent of an area concerned with a concept or issue. It can range from local to global.

sustainability: development that meets the needs of the present without compromising the ability of future generations to meet their own needs.

system: a collection of interrelated components that together form a working unit. In an open system energy and/or material can be gained or lost across the system boundaries. In a closed system there is no movement of energy and/or material across its boundaries.

threshold: a critical level that, if crossed, means a system may undergo sudden change that may not be reversible. The minimum demand or population needed to support a service.

vulnerability: a measure of the extent to which a person or community is likely to be damaged or disrupted by an event.